CAMBRIDGE STUDIES IN PHILOSOPHY

Scientific realism and the
plasticity of mind

CAMBRIDGE STUDIES IN PHILOSOPHY

General editor SYDNEY SHOEMAKER

Advisory editors J. E. J. ALTHAM, SIMON BLACKBURN
GILBERT HARMAN, MARTIN HOLLIS, FRANK JACKSON,
JONATHAN LEAR, JOHN PERRY, T. J. SMILEY, BARRY STROUD

Scientific realism and the plasticity of mind

Paul M. Churchland

Professor of Philosophy, University of California, San Diego

CAMBRIDGE
UNIVERSITY PRESS

CAMBRIDGE UNIVERSITY PRESS
Cambridge, New York, Melbourne, Madrid, Cape Town, Singapore,
São Paulo, Delhi, Dubai, Tokyo, Mexico City

Cambridge University Press
The Edinburgh Building, Cambridge CB2 8RU, UK

Published in the United States of America by
Cambridge University Press, New York

www.cambridge.org
Information on this title: www.cambridge.org/9780521338271

First published 1979
First paperback edition 1986
Reprinted 1987, 1990, 1993, 1995

A catalogue record for this publication is available from the British Library

Congress Cataloging-in-Publication Data is available.

ISBN 978-0-521-22632-5 Hardback
ISBN 978-0-521-33827-1 Paperback

FOR
NOLLAIG

Contents

Preface

This volume is descended from a paper delivered to the Western Division meetings of the Canadian Philosophical Association in 1971. That paper sketched the argument of chapter 2 and the principal thesis of chapter 5. In the interim, several intermediate versions of that material have been presented on a variety of occasions, and I should like to thank the participants, audiences, and departments involved for their kindness and critical suggestions.

The present essay is addressed simultaneously to two distinct audiences. The first audience consists of my professional colleagues, other academics, students, and lay readers, who are less than intimately familiar with the philosophical position commonly called scientific realism. For them I have here attempted to make available in fairly short compass a coherent and comprehensive account of that position as it bears on the philosophy of perception, on the theory of meaning, on the philosophy of mind, and on systematic epistemology. The view proposed is not merely eclectic, however. The synthesis effected is novel in various respects, and the supporting arguments are for the most part novel as well. It is my earnest hope, therefore, that the discussion will be found entertaining, and valuable as well, to those of my colleagues who already share a familiarity with the philosophy of science in general and with scientific realism in particular. This group constitutes the second audience to whom this essay is addressed, and it is my special concern to bring to their attention some generally unappreciated consequences of scientific realism concerning the future directions that epistemological theory and rational methodology must take. These consequences are drawn and explored in the final chapter, and it is there, I think, that the reader will find the most important contribution of this essay.

I should like to express thanks to my teacher, Wilfrid Sellars, for his enormously stimulating influence during and after my years at the University of Pittsburgh. And I must acknowledge a debt of similar magnitude to the writings of Sir Karl Popper, Paul Feyerabend, W. V. Quine, and T. S. Kuhn. Among philosophers of my own generation I must mention first my wife and colleague, Patricia Smith Churchland, whose happy influence defies accounting in less than essay length. And though their instruction has been less sustained, I am pleased to be able to thank Nollaig and Ann Mackenzie, Cliff Hooker, Jay Rosenberg, Charles Morgan, and Bill Harper for a variety of invaluable discussions during sundry visits, encounters, and colloquia. A special acknowledgement must here be paid to Cliff Hooker, who not only had the wit to reach quite independently all or most of the same conclusions reached in chapter 5 (Cf. C. A. Hooker, 'The Philosophical Ramifications of the Information-Processing Approach to the Mind-Brain', *Philosophy and Phenomenological Research*, vol. 36 (1975), but also the kindness to share his enthusiasm and to provide a most timely piece of unsolicited encouragement on what often seemed a hopeless dialectical task.

Thanks also are owed to my departmental colleagues Roy Vincent, Leon Ellsworth, Ken Warmbrod, and Michael Stack for their helpful criticisms and positive suggestions at many points in the writing of this book. Collectively, much advice has been accepted (and disregarded) from all of these philosophers, but since I have accepted only their good advice, the responsibility for residual falsehoods rests entirely with me.

I am grateful also to my brother Mark T. Churchland for much enjoyable discussion of the theoretical examples used in §§3 and 4, and to my friend Kenneth Hughes for his cheerful reading of the final draft.

Turning finally to institutions, I must thank the editors of *Philosophy of Science* for permission to use some material from an earlier publication; the University of Manitoba for a sabbatical leave in which the bulk of this volume was written; and the Canada Council for a leave fellowship and research grant to support the work here presented.

The Sechelt Peninsula   PMC
British Columbia

I
Introduction

1. *The perspective of scientific realism*

The common opinion concerning scientific knowledge and theoretical understanding – of molecules, of stars, of nuclei and electromagnetic waves – is that it is of a kind very different from our knowledge of apples, and tables, and kitchen pots and sand. Whereas theoretical knowledge can be gained only by an act of creative genius, or by diligent study of the genius of another, knowledge of the latter kind can be gained by anyone, by casual observation. Theoretical understanding, it will be said, is artificial where the latter is natural, speculative where the latter is manifest, fluid where the latter is essentially stable, and parasitic where the latter is autonomous.

That these specious contrasts are wholesale nonsense has not prevented them finding expression and approval in the bulk of this century's philosophical literature. Theoretical "knowledge" is there represented as an essentially peripheral superstructure erected on the body of human knowledge proper. This approach did promise some advantages. One could hope to give an account of the semantics of theoretical concepts by explicating the special kinds of relations they must bear to non-theoretical concepts; and one could hope to give an account of the warrant or justification of theoretical beliefs by explicating the relations they must bear to our non-theoretical knowledge. That is, taking the non-theoretical as a temporary given, one could hope to provide a successful account of theoretical understanding *short of* the larger business of constructing an account of human understanding in general.

This now appears to be impossible. The failure of the various accounts proposed forms part of the reason, but the main consideration is rather more interesting. The premiss on which the older approach is based – that there is indeed a genuine distinction between the theoretical and the non-theoretical – appears to be false. Upon close inspection the various contrasts thought to fund the distinction are seen to disappear. If viewed warily, the network

of principles and assumptions constitutive of our common-sense
conceptual framework can be seen to be as speculative and as
artificial as any overtly theoretical system. It even displays fluidity,
if one takes a decently long view. Comprehensive theories, on the
other hand, prove not to be essentially parasitic, but to be potentially autonomous frameworks in their own right. In short, it
appears that all knowledge (even perceptual knowledge) is theoretical; that there is no such thing as *non*-theoretical understanding.
Our common-sense conceptual framework stands unmasked as
being itself a theory, or a battery of related theories. And where
before we saw a dichotomy between the theoretical and the non-
theoretical, we are left with little more than a distinction between
freshly minted theory and thoroughly thumb-worn theory whose
cultural assimilation is complete.

We cannot, therefore, adopt an instrumentalist or other non-
realist attitude towards the doctrines and ontologies of novel
theoretical frameworks, unless we are prepared to give up talk of
truth, falsity, and real existence right across the board. And a
further corollary is that our common-sense framework must
acknowledge its vulnerability to the same sorts of criticism that
decide theoretical questions generally.

In particular, we are forced to acknowledge the demise of the
familiar distinction between theoretical beliefs and perceptual
beliefs. The class of perceptual beliefs must now be counted as a
subclass of the class of theoretical beliefs: roughly, as those singular
theoretical beliefs acquired as spontaneous non-inferential responses
to sensory states of the perceiver. As with singular theoretical
judgements generally then, the adequacy of our perceptual judgements is in part a matter of the adequacy of the background theory
(conceptual framework) in whose terms they happen to be framed.
Our perceptual judgements can no longer be assigned any privileged
status as independent and theory-neutral arbiters of what there is
in the world. Excellence of theory emerges as the fundamental
measure of all ontology. The function of science, therefore, is to
provide us with a superior and (in the long run) perhaps profoundly different conception of the world, *even at the perceptual
level*.

This last position, one of the central theses of scientific realism,
is commonly indicted on grounds that it throws objectivity to
the winds. Thus framed, this is an insensitive criticism. The

manner in which the varieties of empiricism provide for objectivity (comparison of theory with some stable and theoretically unbiased level of perceptual knowledge) is at the same time an analysis of what objectivity *is*. If the manner of provision is suspect, then so is the analysis. That scientific realism renders objectivity *thus construed* impossible is therefore a question-begging complaint at best.

The true situation here wants emphasizing. We have on the one hand the classical empiricist picture: a special class of singular judgements for which failure of objectivity is impossible, general failure at least, and in which the objectivity of everything else is grounded. But empiricists have never succeeded in providing a satisfactory account of the "grounding" relation, nor, more importantly, of the autonomous justification of the "foundational" judgements. Failure on the first count is easily tolerated, perhaps, for the matter is complex and the failure has not been total. Failure on the second count, however, is less easily tolerated, for here the failure has been complete. And when this division of labour has itself become suspect – by way of the apparent theory-ladenness of evidence and our inability even to isolate the neutral foundation, let alone account for it – then it is clearly time to be looking at other approaches, undeterred by the argument from frying pans and fires.

Still, it must be conceded that the methodological problem – the problem of reckoning just what the objectivity/integrity/rationality of the intellectual process consists in – remains wide open, and there is not the slightest sign of a gathering consensus even among (especially among) those sympathetic to scientific realism.[1] One school of thought continues to insist on or to hold out for unique normative principles to help us steer an unquestionably rational course between the dogmatism of conventionalism and the anarchy of a nihilistic pluralism. Another strand argues that a sociology or a psychology of knowledge is the best we can hope for in this matter. The first approach has been described by some as quixotic, and the second by others as irrelevant. For my own part, I am not prepared to concede that normative epistemology is impossible. But I also harbour an irreversible sympathy with those who recommend that epistemology be "naturalized". More generally, I suggest that the

[1] See, for example, the papers collected in *Criticism and the Growth of Knowledge*, ed. Lakatos and Musgrave (Cambridge, 1970).

methodological problem will not be solved short of an intellectual revolution in our conception of ourselves as intellectual beings. Why this is so, what kind of revolution is required, and what sort of solution it might provide will be discussed in chapter 5.

I have introduced this essay with a problem, and I shall close it with an extended discussion of that problem. But in between there is much to be said. Scientific realism is a remarkably fertile philosophical position, and it is the other major purpose of this essay to illustrate and exploit that fertility in a variety of philosophical topics. First among these is the matter of perception. In chapter 2 I shall argue for the theoretical character of perceptual judgements by a dialectical route that will highlight the possibility of our being trained to make systematic perceptual judgements in terms of theories *other* than the common-sense theory we learned at mother's knee. This possibility has been mentioned or scouted briefly by a number of philosophers in the tradition at issue, but it has never been explored in any detail. I hope to remedy this, to some extent. Past and present science provide us with a variety of alternative conceptions of this or that familiar perceptual domain, and the displacement, at the level of spontaneous perception, of some of our current conceptions is readily imagined in considerable detail. There are, I think, some surprises waiting for us here.

In addition to its intrinsic interest, that discussion will raise broader issues concerning the role of perception in the evolution of human understanding, the status of the familiar ontology of common sense, and the matters of cross-theoretical comparison and intertheoretic reduction. These raise in turn the more general issues of meaning, translation, and conceptual change, and chapter 3 is devoted primarily to these. A quasi-holistic theory of meaning will there be outlined and defended against the most prominent of the opposing views. In particular, Quine's rejection of the analytic/synthetic distinction is there endorsed and amplified, but his positive views on meaning and translation are themselves rejected in turn. The semantic theory that emerges from this discussion is then put to work in an attempt to throw some light on the nature and significance of incommensurability, on the nature of cross-theoretical comparison, and on the nature of intertheoretic reduction.

All of this, it seems to me, reaches a subclimax in the area of the philosophy of mind: in the other-minds problem, in the problematic epistemology of self-knowledge, in the curious features allegedly unique to mental states, and in the mind/body problem. The revolutionary idea that our conception of *ourselves* as thinkers/ perceivers/desirers is *also* a thoroughly theoretical conception places the problems listed in a most intriguing and profitable perspective. Wilfrid Sellars' 'Empiricism and the Philosophy of Mind'[2] is the landmark paper here, but much more needs to be said. Included in the rich fallout of this idea is, for example, the possibility that our self-conception, our common-sense "theory of persons", is false.[3] On the matter of the traditional mind/body problem, this possibility is represented by a position that has come to be called 'eliminative materialism' (sometimes called 'the disappearance theory of mind'). Its immediate adversary is *reductive* materialism (more commonly called 'the identity theory of mind'). The essential issue between these positions is whether our common-sense theory of persons will prove reducible to the projected neurophysiological account of human behaviour that exists in almost everyone's utopian vision of our scientific future. The eliminative materialist will argue (1) that there is no reason why it must prove so reducible; that there is no reason to make it a constraint, for example, on the adequacy of a general materialistic theory of *homo sapiens* that it be capable of reducing our common-sense theory of persons, and (2) that there is some reason for doubting that these two theories will in fact form a reduction pair. According to him, then, the prospect we face is that a detailed neurophysiological conception of ourselves might simply displace our mentalistic self-conception in much the same way that oxidation theory (and modern chemistry generally) simply displaced the older phlogiston theory of matter transformation. That we are long in the habit of making non-inferential introspective judgements in the terms of the theory to be displaced affects the matter not at all. Those more primitive habits can also be displaced, and we may look forward to framing our introspective judgements directly in terms of a much more adequate and powerful theory of human psychology.

So far, this is all part of the subclimax. From the point of view

---

[2] *Minnesota Studies in the Philosophy of Science,* vol. 1., ed. Feigl and Scriven (Minneapolis, 1956); reprinted in Wilfrid Sellars, *Science, Perception, and Reality* (London, 1963).

[3] See Paul Feyerabend, 'Materialism and the Mind–Body Problem', *Review of Metaphysics,* 17 (1963), and 'Mental Events and the Brain', *Journal of Philosophy,* vol. 60, no. 11 (May 1963). See also Richard Rorty, 'Mind–Body Identity, Privacy, and Categories', *Review of Metaphysics,* vol. 1 (1965).

of the present essay, the real climax occurs when we pursue these
considerations into epistemology. As epistemology is currently
conducted, the questions and answers are posed and expressed
within the framework of the common-sense theory of persons. Our
concern has standardly been with the *rationality* of *belief,* or of sets
of beliefs, or of transitions from one set to another. And we have
attempted to characterize the essence of epistemic virtue in terms
of the properties and relations holding of and among beliefs, or
more characteristically, among the sentences that express them. But
if the common-sense theoretical conception that provides the
parameters of this approach is itself profoundly awry, or even
seriously superficial, then there is no good reason to expect that
these "sentential" epistemologies will ever achieve anything more
than scattered and partial successes. And in fact, as the honest
among us must admit, that is all they have ever achieved.

The perspective of scientific realism here allows us to suggest
a diagnosis of the current impasse in epistemology, of the seeming
intractability of the methodological problem. The diagnosis, in a
nutshell, is that in pursuing a *kinematics of sentences* we have been
concerned with parameters that are idiosyncratic to our own species
and culture, parameters that reflect only very superficially what is
really going on. Once this possibility has been clearly perceived,
there emerge a number of considerations that indicate that this is
no mere possibility, but a near certainty. These considerations will
be detailed and discussed in chapter 5. The upshot of the discussion
is that we must contrive to step out of our parochial self-conception,
to transform our narrow concern with the rationality of belief into
a global concern with the parameters of operation of "epistemic
engines" generally. An exploration of what this might involve will
concern the final section of this essay.

2

The plasticity of perception

2. *The semantics of observation predicates*

The guiding conviction of this chapter is as follows: *perception consists in the conceptual exploitation of the natural information contained in our sensations or sensory states.* This view will be articulated more fully as we proceed, but even in this rough formulation it suggests a question: how efficient are we at exploiting this information? The answer, I shall argue, is that we are not very efficient at it, or rather, not nearly as efficient as we might be. What needs to be made clear in the topic of perception is the truly vast amount of exploitable information, contained in our own sensations, that goes blissfully *un*exploited by our conceptually benighted selves.

As indicated earlier, this myopia appears remediable. Our current modes of conceptual exploitation are rooted, in substantial measure, not in the nature of our perceptual environment, nor in the innate features of our psychology, but rather in the structure and content of our common language, and in the process by which each child acquires the normal use of that language. By this process each of us grows into a conformity with the current conceptual template. In large measure we *learn*, from others, to perceive the world as everyone else perceives it. But if this is so, then we might have learned, and may yet learn, to conceive/perceive the world in ways other than those supplied by our present culture. After all, our current conceptual framework is just the latest stage in the long evolutionary process that produced it, and we may examine with profit the possibility that perception might take place within the matrix of a different and more powerful conceptual framework.

The obvious candidate here is the conceptual framework of modern physical theory – of physics, chemistry, and their many satellite sciences. That the conceptual framework of these sciences is immensely powerful is beyond argument, and its credentials as a systematic representation of reality are unparalleled. It must be a dull man indeed whose appetite will not be whet by the possibility of perceiving the world directly in its terms.

Before we attempt to anticipate the millennium, however, it

needs establishing that this *is* a possibility. To do this I must try to construct a framework of intuitions to compete with (and I hope supplant) the crude intuitions supplied by common sense. The key theses here to be argued concern (1) the semantics of observation predicates, and (2) the intentionality (the "of-ness") of sensations. A proper perspective on both these matters can be had, I believe, by examining a case in which our ordinary observation vocabulary for temperature is used by beings with sense organs different from those which enable us to use that vocabulary.

As we are now constituted, we lack the sensory equipment to perceive *visually* the middle-range temperatures of common objects. But it is not difficult to imagine beings who could. Simply imagine a race of men with larger eyeballs and/or more highly refractive lenses, a race of men whose retinas consist solely of rods sensitive to electromagnetic radiation at some wavelength in the far infrared. Since the vigour with which any body radiates in the far infrared is a more or less straightforward function of its temperature, and since images of these bodies will be formed on the retinas of the kind of eyes described, their possessors will be quite prepared, physiologically, to perceive visually the temperatures of common bodies, since the "brightness" of the corresponding image will be a function thereof.*

Let us imagine then an independent society of such beings speaking a language that is, superficially at least, indistinguishable from English, save for two points. First, it lacks our colour vocabulary, including 'black', 'grey', and 'white'. And second, our ordinary temperature vocabulary ('cold', 'hot', 'warmer than', etc.) is learned by the very young as an observation vocabulary for *visual* instead of tactile reports. (To simplify things later on, assume also that these beings lack any tactile or bodily sense for temperature, as we lack any tactile or bodily sense for colour.) They acquire the use of the relevant predicates in much the way we learn the use of our colour predicates, and they steadily amass, in various ways, the usual set of general beliefs or assumptions concerning temperature: 'Fires are hot', 'A warm thing will warm up a cooler thing, but never the reverse', 'If a body is warmer than a second body, and that second body is warmer than a third body, then the first body is warmer than the third body', 'Food keeps better in a cold place',

---

* To work at all well, such eyes would have to be cooled by some means to minimize the background noise of their own thermal radiation. Let it be so. My thanks to Richard Gregory for catching this lapse.

'Hot things cause painful burns', 'Rubbing things makes them warmer', and so on and so forth. Generally speaking, the set of "temperature beliefs" of any adult member of this imagined society is no more dissimilar to your own set than yours is to, say, your next-door neighbour's. A few differences are pretty much standard, of course. 'Temperatures can be seen' is a platitude for them, but will be counted false or problematic by most of us. And the reverse holds for 'Temperatures can be felt'. But given the differences in our sensory equipment and the lack of pressure on the common man to consider the possibility of other sensory modalities, such occasional failures of perfect correspondence are neither surprising nor very interesting.

Given their linguistic behaviour, the special nature of their eyes, and the accuracy *de facto* of their perceptual reports on temperature, the natural position to take is that these people can indeed visually perceive the temperatures of objects, at least under "normal" conditions (that is, under conditions of relatively low ambient infrared radiation, conditions such that the infrared that bodies *emit* is not swamped behind the infrared they may *reflect* to us from other sources). As they see it, the visually perceivable world consists not of middle-sized and variously coloured material objects, but rather of middle-sized and variously *heated* material objects. That is, they perceive hot objects *as* hot (warm, cold); they can visually perceive *that* they are hot (warm, cold).

If we accept this conclusion concerning their perceptual capabilities, we should notice that we have done so without the benefit of any information concerning the intrinsic qualities of their visual sensations. Should this affect the matter? To make matters interesting, let us suppose finally that, so far as the intrinsic nature of their visual sensations is concerned, the world "looks" to them much as it looks to us in black-and-white prints of pictures taken with infrared-sensitive film. (This is in any case the result to be expected, since (a) their retinas contain only rods, and (b) we are supposing their physiology to be entirely human beyond the peripheral respects cited.) That is to say, on viewing a very hot object they have what *we* would describe as a sensation of an incandescent *white* object, and on viewing a very cold object they have what we would describe as as sensation of a *black* object, and so on. They, of course, describe these sensations quite differently – as sensations of heat, of coldness, and so on.

With these assumptions we arrive at the heart of the matter. If we succumb to the common-sense view that the meaning of simple observation terms is given in sensation, we must insist that *their* terms, 'cold', 'warm', and 'hot' really mean *black, grey,* and *white* respectively, rather than *cold, warm,* and *hot.* But this heterophonic sensation-guided translation of their vocabulary is not without very serious consequences. If we adopt it, we shall have to count as false all (but for a few flukes) of their many background beliefs involving the predicates in question: try substituting 'black', 'grey', and 'white' respectively for 'cold', 'warm', and 'hot' in the sample temperature beliefs listed above. Equally bad, we shall have to count as false all of their "observation" judgements involving the relevant terms (save for those accidental cases where, for example, a cold object just happens to be black), for they certainly cannot see whether objects *are* black, grey, or white. Their visual sensations are keyed to other parameters entirely.

Accordingly, to insist on this sensation-guided translation is, I suggest, to make a joke of a perfectly respectable and very powerful sensory modality, and of a simple and appropriate mode of conceptual exploitation which has every virtue we can claim for our own habits of judgement in matters visual. While it is true that, under normal conditions, *this* is how white things and only white things look to standard observers, it is equally true that under (different) normal conditions, *this* is how *hot* things and only hot things look to (different) standard observers. To regard the relevant class of visual sensations as uniquely appropriate only for our kinds of visual judgements is to be insupportably parochial.

And we are not limited here to pleading for sympathetic intuitions. The crucial consideration is the following. If we insist, by way of this sensation-guided heterophonic translation, on thus making a joke of their beliefs and visual capabilities, we must be prepared to have the very same joke made of our own beliefs and visual capabilities with respect to black, grey, and white. The fact of the matter is that their beliefs, background and observational, involving their 'cold', 'warm', and 'hot', do not match up at all with our beliefs, background and observational, involving our 'black', 'grey', and 'white'; and the failure of match is such that if we insist that these two predicate families are genuine translational correlates even so, then we must insist that at least one of these two sets of beliefs is systematically false. But *which?* If our infrared cousins adopt the same strategy of translation, as by parity of reasoning they must, they will be as eager and as well prepared to insist that it is *our* beliefs, background and observational, that must be discounted as systematically false. As they translate us, they will be compelled to regard our observation judgements on black/grey/white things as hopelessly wide of the mark; and in translating our background beliefs, they will find in us silly convictions like 'Snow is hot', 'Africans are colder than Europeans', 'A hot shirt shows the dirt more readily than a cold one', and so forth. Their story on us is the image of our story on them; it is as uncomplimentary, and it is as stupid.

In short, the sensation-guided translation lands us in an epistemological dilemma to which there can be no resolution. Both their story and ours must be rejected as thoughtless and parochial, for there is no relevant asymmetry between the respective cases to sustain the one piece of foolishness while unmasking the other. Translating their 'cold' as our 'black', and so on, is therefore out of the question. Quite aside from the unwarranted nonsense that translation would make of their background beliefs and perceptual capabilities, there is nothing to locate the relevant brand of nonsense in their case rather than in our own.

By contrast, the straightforward homophonic translation of their temperature vocabulary has every empirical virtue it is possible for a translation to have. Our general background beliefs involving the relevant terms all match up; there is systematic agreement between us on specific cases (that is, the relevant extensions all match up); where our sense modalities overlap (and substantially they do) all of our observation judgements match up; causal analysis of our respective modalities reveals (confirms) that they are both responding to the same external parameter: temperature. The residual difference between us is just that we have different means of detecting this feature.

The impossibility of the heterophonic translation discussed earlier is just the impossibility of the thesis that the meaning of the common observation terms at issue is given in sensation. And the independent appeal of the homophonic translation is just the appeal of the view that the meaning of the relevant observation terms has nothing to do with the intrinsic qualitative identity of whatever sensations just happen to prompt their non-inferential application in singular empirical judgements. Rather, their position in semantic space appears to be determined by the network of sentences containing them accepted by the speakers who use them. This would return us to our original conclusion: of course the beings described can visually perceive the temperatures of material objects. They can even say so, and mean it.

It may be objected that there is a *tertium quid* here, that the wholesale adoption of the homophonic translation is not the only alternative to the particular heterophonic translation rejected above. Perhaps a *part* of the meaning of the relevant terms is given in sensation, while the remainder is fixed by a cluster of background beliefs. A proponent of this view could concede that, since the infrared people's "temperature" terms are *extensionally* equivalent to our temperature terms, the homophonic translation might be "adopted" (with a broad wink) for practical purposes, while still insisting that our own language contains *no* genuine translational correlates for the alien terms at issue (since there are no terms in our language that are both prompted by the relevant sensations and figure in a set of sentences accepted by us that matches the set accepted by the aliens).

But this third alternative has nothing to recommend it over the homophonic alternative – quite the reverse. It requires us to deny that the beings with the infrared eyes can perceive the temperatures of objects, and indeed to deny that *any* beings, no matter what their sensory apparatus, can perceive the temperatures of objects unless they are subject to precisely the same range of bodily sensations with which we *happen* to respond to hot and cold objects. If they do not have these sensations, they cannot even have the concept of temperature, let alone perceive whether and to what degree it is instanced in objects; they must be perceiving something else unknown to us.

If there is any plausibility to this sort of position, it derives, I suggest, from the contingent fact that relative to us both temperature and colour are monomodal perceptual properties. To help free our intuitions, let us consider some properties that are bimodal for us. For example, small objects can be felt to be round, and they can also be seen to be round. Consider now two people, one of whom can only see, while the other can only feel, whether objects are round. The latter is congenitally blind, say, while the former is sighted but congenitally paralysed or benumbed. Shall we say of them that they can both perceive whether objects are round, though by way of different sensory modalities? Or shall we insist that they must be perceiving different properties, that the ordinary term 'round' is systematically ambiguous, and that their respective 'rounds's are intertranslatable only with a broad wink? I take it that the former is the plausible alternative. But if it is, then we must ask how the case of the infrared people and ourselves, with respect to 'warm', differs from the case of these two differently equipped humans with respect to 'round'. To draw the relevant parallel differently, let us assume that our infrared cousins have the normal tactile or bodily sense for temperature after all. Could they not then both feel and see that objects are warm, just as we can both feel and see that objects are round? Would not temperature then be a bimodal feature for them, just as roundness is for us?

When we reflect on considerations such as these, it is tempting to think that, had our sensory equipment been sufficiently rich that all properties observational for us were at least *bi*modal for us, the idea of a distinct kind of phenomenological *meaning* would never have been conceived in the first place.

If it is possible for other beings to share with us a common observation vocabulary – our present vocabulary, for example – despite differences in their sense organs and sensations, then the view that the meaning of our common observation terms is given in, or determined by, sensation must be rejected outright, and as we saw, we are left with networks of belief as the bearers or determinants of understanding. This possibility has been illustrated only for the case of our ordinary temperature vocabulary, but the lesson here is general. For any observational property φ conceived by us to be an objective feature of the world, we must then concede the possibility that φ be sensorily detected by means other than the particular means used by us. Consider then a being suitably (but differently) equipped to detect the occurrence of φ. He may respond with a term 'θ' where we use 'φ', but the sentences to which he assents, both singular and general, involving 'θ', systematically match or map onto the sentences we accept, both singular and general, involving 'φ'. In such a case, 'φ' is clearly an ideal translation for 'θ', and any differences in the sensations that respectively prompt their non-inferential application do not amount to a row of pins so far as the propriety of that translation is concerned.

What all this illustrates is the necessity of distinguishing between our *understanding* of even the simplest of our observation predicates, and *our ability to apply them non-inferentially* in response to whatever sensations (if any) nature happens to have given us. An infant, trained this morning to respond to red objects with 'red', has the latter without the former; and a blind man can have the former without the latter. Lacking a suitable sensational response to coloured objects, a blind man cannot recognize colours observationally. But this is an epistemic rather than a semantic failing. Conversely, a child's initial (stimulus–response) use of, say, 'white', in response to the familiar kind of sensation, provides that term with no semantic identity. It acquires a semantic identity as, and only as, it comes to figure in a network of beliefs and a correlative pattern of inferences. Depending on what that acquired network happens to be, that term could come to mean *white*, or *hot* (as we saw), or any of an infinity of other things.

None of this should occasion surprise. If a term is to occupy a determinate position in semantic space, then it must be *connected* in that space; it must stand in semantic relations to other elements in semantic space, and its position will be determined by these. Other *terms* count as such elements, but sensations themselves do not.

3. *The conceptual exploitation of sensory information*

The argument of the preceding section – we might call it 'the argument from transposed modalities' – heralds the need to distinguish between two kinds of intentionality, between two senses in which a given sensation can be a "sensation of φ".

Objective intentionality:

  A given (kind of) sensation is a sensation of$_o$ φ with respect to a being x if and only if

  under normal conditions, sensations of that kind occur in x only if something in x's perceptual environment is indeed φ.

Subjective intentionality:

  A given (kind of) sensation is a sensation of$_s$ φ with respect to a being x if and only if

  under normal conditions, x's characteristic non-inferential response to any sensation of that kind is some judgement to the effect that something or other is φ.

These formulations are rougher than they should be, but they will serve my present purposes. The first definition is meant to capture one sense in which our sensations contain information concerning our environment. The second definition is meant to reflect a sense in which we (try to) exploit such information. It is clear from the above that neither the objective nor the subjective intentionality of a given kind of sensation is an intrinsic feature of that kind of sensation. Rather, they are both relational features, involving the sensation's typical causes in the former case, and its typical (conceptual) effects in the latter. And it is equally clear that both the "of$_o$-ness" and the "of$_s$-ness" of one and the same kind of sensation can vary from being to being, and even over time within the history of a single individual, the variation being a function of differences or changes in sensory apparatus in the case of objective intentionality, and of differences or changes in training and education in the case of subjective intentionality.

In terms of this distinction, the lesson of the preceding section is threefold. First, what one is physiologically capable of perceiving is solely a matter of the objective intentionality of one's sensations, of the actual information they naturally contain. Second, what properties one actually does perceive the world as displaying is additionally a matter of the subjective intentionality of one's sensations, and of whether this matches their objective intentionality. And third, the intrinsic qualitative identity of one's sensations is irrelevant to what properties one can or does perceive the world as displaying. The meaning of a term (or the identity of a concept) is not determined by the intrinsic quality of whatever sensation happens to prompt its observational use, but by the network of assumptions/beliefs/principles in which it figures. Sensations are just *causal* middle-men in the process of perception, and one kind will serve as well as another so long as it enjoys the right causal connections. (So far then, in principle they might even be *dispensed* with, so far as the business of learning and theorizing about the world is concerned. As long as there remain systematic causal connections between kinds of states of affairs and kinds of singular judgements, the evaluation of theories can continue to take place. This I take to be the point of Paul Feyerabend's cryptic paper, 'Science Without Experience', *Journal of Philosophy*, vol. 66, no. 22 (1969).)

If all this is correct, the possibility of a dramatic modification and expansion of the domain of human perceptual consciousness – without modification of our sense organs – becomes quite real.

The reason is simple. The objective intentionality of a kind of sensation consists in its being a reliable empirical indicator of the presence or value of some environmental feature or parameter, and in this sense the various kinds of sensations normal to human beings are, as a matter of empirical fact, sensations of$_0$ a great deal more than our current modes of conceptual exploitation would suggest. The raw material is therefore already there, waiting to be exploited by some more educated patterns of conceptual response on our part. Before pursuing this possibility directly, however, we must come to a more detailed understanding of the nature of those conceptual frameworks which make possible the conceptual exploitation of sensory information. In this regard it will prove instructive to examine what we can all agree is a case of *mis*exploitation.

Suppose again an isolated society of humans whose physiology this time differs in no way from our own: same sense organs, same sensations, same objective intentionalities. Let their language be indistinguishable from ordinary English, save for the following. Where we think and speak of bodies as being warm, as growing cold, as glowing with heat, and so on, they think and speak quite differently. As they conceive of things, and have conceived of things as long as their history records, all material bodies – solid, liquid, and gaseous – contain an exceedingly subtle compressible fluid called *caloric*. Caloric is held in material bodies, in varying amounts, rather in the way water is held in sponges. Caloric flows or diffuses itself equally throughout any uniform body, and will flow from one body to another if the two bodies are in physical contact and the caloric fluid pressure in one exceeds the caloric fluid pressure in the other. Rather like water, one might say, caloric "seeks its own level".

Substances differ, however, in their affinity for, or their capacity to soak up this fluid. That is, some substances are better "sponges" than others. Certain substances will happily absorb large amounts of caloric fluid while the pressure of the absorbed fluid rises only slightly. But in other substances the pressure of the absorbed fluid rises sharply with only small additions of fluid from the outside, and it quickly reaches an equilibrium with the pressure of the donating body, wherein the flow ceases. Water, for example, is an excellent caloric sponge, while aluminium is a relatively poor one. Finally, of course, substances differ in the rate at which caloric flows through them.

As these people blithely explain to us, caloric is, at common pressures, perfectly transparent (invisible), but it is easily visible at very high pressures when it first becomes a highly distinctive red, then orange, yellow, and finally white as it is subjected to further increases in pressure. It is easily *felt*, however, even at common pressures. Skin contact with bodies at various caloric pressures produces characteristic sensations as the fluid flows into (or out of) the observer's body. That is, these people claim to *perceive* or *observe*, by feeling and on occasion by looking, that material bodies contain caloric fluid at various pressures, and even to perceive the fluid itself, as when they feel it flowing into their fingertips when they touch a body at high caloric fluid pressure, or see it in a body at very high caloric fluid pressure. The children of this society, we discover, are taught from the beginning to use the relevant vocabulary in making observation reports. In those situations where we learned 'is hot' and 'is cold', they learn to use 'is at high caloric pressure', and 'is at low caloric pressure'. And this initial verbal behaviour is quickly augmented and refined by further training so as to reflect their parents' common-sense beliefs about caloric and its behaviour, beliefs which are in fact easily illustrated even to a child, using no more "apparatus" than a source of heat, a small variety of substances, and his own two hands.

The reader will recognize in all this a slightly tarted up version of a now defunct theory of heat. But the suggestion that there is anything theoretical about caloric is met with bewilderment and derision by the good folk just described. None of their contingent beliefs about the nature and behaviour of caloric, they insist, are anything more than simple generalizations of direct experience. And the claim that there is really no such thing as caloric is met with blank stares: they can feel the stuff and even see it. Granted, they concede, what caloric fluid really is in its inner or microscopic nature is indeed a matter for theoretical speculation and discovery, but that is another matter entirely. Whatever caloric fluid is, they insist, it manifestly exists!

While rejecting their position, we must also try to appreciate it. Their conception of things is a fairly powerful one, much more useful and discriminating than our own feeble (common-sense) conception of 'hot' and 'cold'. They will notice, expect, exploit, and explain as a matter of course common phenomena which, if we notice them at all, are quite mystifying. Why does a body expand in proportion to the amount it is heated? To one who conceives of such bodies as confining a fluid under increasing pressure, the phenomenon is only natural. Why do contiguous bodies at different temperatures always end up at one and the same temperature, a temperature somewhere between the initial extremes? To one who conceives of these "temperatures" as unequal pressures in two connected fluid reservoirs, an exchange of fluid until equilibrium is reached is the inevitable result. And why is that equilibrium point often so much closer to one of the initial extremes than to the other? To one familiar with the different caloric capacities (the "sponge factor") characteristic of different substances, the phenomenon is only to be expected. More generally, little or nothing in their common experience will do anything but further encourage and entrench this very natural conception of things. No doubt the more curious among them will have some unanswered questions about the nature and behaviour of caloric, but no more than they have unanswered questions about the nature and behaviour of physical objects. Their uncritical confidence in their "observation" judgements is entirely understandable.

How then shall we convince them that they do not perceive or observe what they suppose, that their conception of the relevant phenomena is confused and mistaken? An appealing approach is that embodied in the following rhetorical question:

> Can you really just *feel* that there is in this object a *fluid*, a *substance* that *flows* and is under *pressure*?

But our friends, boasting a philosophical tradition of their own, have been through all this before. They retort:

> Can you really just *see* that there is a red physical object in front of you? Can you *see* that it has *substance* behind its facing surface, *resistance* to penetration by other bodies, non-zero *heft*, and any of the other features so essential to being a physical object? Can you *see* that it would also look red to *other* observers?

There is no satisfaction to be had from pursuing this approach. Their conception of things is no *more* subject or vulnerable to general epistemological scepticism than is our own conception of things. And as far as localized scepticism is concerned, as in MacBeth's 'Is this a [reservoir of high pressure caloric] I [feel] before me?', they are as skilled in settling such questions as are we.

For example, where we poke a suspect physical object with another (observably) physical object to determine (feel) if it offers resistance, they place a putatively calorified body in contact with another (observably) less calorified body to determine (feel) if their caloric pressures equalize in the way they should.

We may seek solace in the fact that their claims will not withstand all possible scrutiny. After all, their common-sense theory is shown to be false, if not by some flat incompatibility with "observable facts", then by its explanatory impotence in certain cases, by the superior virtues of the very different corpuscular/kinetic theory of heat, and by its lack of any coherent and non-redundant place in the broader conception of the world provided by modern physical theory. Perhaps we can undermine their tenacious faith in their idiom by the simple expedient of showing them that their observation claims and the background beliefs they presuppose are uniformly *false*.

Let us bring them up to date then on the theory of matter. Leaving nothing to chance, we teach them the whole story as we know it, starting with molecular chemistry, through classical and quantum atomic physics, up to and including quantum field theory. And in particular, we teach them what we shall have to call, in deference to their views, 'the corpuscular/kinetic theory of *calorific* phenomena'. What they conceive as "the amount of caloric in a body" is shown to be just the summed energy of motion of the agitated particles that make up the body. What they conceive as "the pressure of the contained caloric" is shown to be just the mean of the various individual kinetic energies of the individual particles. "Flow of caloric fluid" is shown to be just the spread of particle agitation. The "colour of caloric" is displaced by the radiative behaviour of the body itself. What was intelligible in their terms becomes newly intelligible to them, and many mysteries become intelligible for the first time: for example, the "appearance of caloric" when motion meets frictional resistance, the "disappearance of caloric" in motion-producing heat engines, and the "release or production of caloric" in any fire. In short, we convince them that the corpuscular/kinetic theory is *true*.

But their reaction is surprising. They do not concede that their "observation" judgements concerning caloric were (or are) unwarranted, nor that they are false. Instead, they express gratitude for having been enlightened on what caloric fluid really is in its inner nature, and on why it has the observable properties it has: pressure, flow, colour, weightlessness, and so on. The claim that there is really no such thing as caloric meets the same blank stares it met before.

This residual failure to understand, though substantial, may appear easily rectified. How, we ask them, can the collective *agitation* of a bunch of particles *be* a fluid *substance?* And how could a substance possibly be "converted" into *motion* (as they will have to suppose happens when the agitated particles of a confined gas give over some of their energy to the large piston they force into motion)? Conversely, how could motion possibly be "converted" into a *substance* (as they will have to suppose happens when the motion of a hammer is given over to micro-motions of the particles of the anvil it strikes)? Aside from the particles of the bodies involved, we explain, there is no substance here at all, weightless or otherwise. Only motions, concerted and chaotic.

This finally produces some confusion, but not capitulation. Some concede that if the kinetic theory is true then caloric is a very strange sort of substance indeed, but they insist that the beliefworthiness of the kinetic theory *derives* from its ability to explain the observable behaviour of *caloric*, and no acceptable theory can coherently deny its own evidential base. Others mutter darkly about "emergent substances". Some invent Instrumentalism, and resolve the conundrum by consigning the kinetic theory to the class of useful fictions or truthvalueless calculation devices. And still others pose us an embarrassing question. If we take the truth of the kinetic theory as settling the matter of the non-existence of caloric, how is it we are so sanguine about the existence of material objects given the truth of special relativity and the quantum field theory? How, they parody, can nothing more than a harmonious mosaic of standing and travelling *waves* in various *fields* in otherwise empty space *be* a material *substance?* And how can a substance possibly be "converted" into energy, and in particular, into *motion?* Do not the theories cited plus the empirical fact of nuclear explosions force on us the idea that substances can and do change into motion? Overall then, are not our own perceptual judgements about material objects in the same dubious position as their "perceptual" judgements about caloric? Whence this highly selective prejudice against caloric?!

The point of the preceding is not that history has misjudged caloric. But one can appreciate that history's (accurate) judgement against the conceptual framework of caloric theory was made much easier by the fact that said framework was *not* the entrenched medium of everyman's perceptual judgement. What might be called 'conceptual inertia' is at a maximum for those concepts customarily applied in perception, and that inertia forms an unfortunate barrier to the accurate appreciation of theoretical issues when those issues happen to reflect back on the propriety of one's own conceptual habits in observation. We have just seen this resistance hard at work. What our calorified friends must be helped to appreciate is that the conceptual network in whose terms they are so accustomed to framing their perceptual judgements is itself a theory, one theory out of an infinity of others any one of which might have enjoyed the same place in their affections. They will then be able to appreciate the details of their situation for what they are. First, the caloric framework must be evaluated *as* a theory, by comparing its virtues with those of alternative theories. Second, the fact that they are profoundly accustomed to thinking and perceiving in its terms is historically accidental, culturally idiosyncratic, and epistemologically irrelevant. Third, their common-sense theoretical framework has been superseded by a vastly superior theory, the corpuscular/kinetic theory, a theory to which the caloric theory is reducible only *very* grossly, if at all. And fourth, there is therefore no *reason* for them to exercise themselves so desperately in defence of the ontology of the old theory and the propriety of their customary patterns of perceptual judgement. This is not to say they should be ashamed of either – they could have done much worse than embrace the caloric theory – but the time for serious conceptual change for them has clearly come.

Having gained some understanding, at one remove from ourselves, of what the conceptual exploitation (or misexploitation) of sensory information can involve, it is time to ask if the lessons just urged on our calorified cousins should not be addressed with equal urgency to ourselves. We must ask how, if at all, our own exploitative situation differs from theirs. The initial presumption must be that it does not, for we saw in §2 that the meaning of common observation terms like 'hot' and 'cold' is determined by the cluster of beliefs and assumptions in which they figure. In this respect they are on all fours with overtly theoretical terms whose meaning is grasped by way of an appreciation of the theory (the set of statements) which introduces them. That the common terms at issue happen to enjoy an observational role is, as we saw in §2, semantically irrelevant. From a purely semantic point of view then, our common observational framework for temperature is indistinguishable from a theoretical framework.

The similitude is equally plain when viewed from the standpoint of explanatory power. To see this clearly we need only imagine a case (a complement to the fable in §2) wherein the observational role of our common-sense temperature concepts is simply subtracted, leaving their remaining functions intact. Conjure up an isolated society of English speakers again, in whom we find all of our usual temperature beliefs plus agreement on specific cases, but who turn out to lack *any* sensory equipment for temperature. They are "temperature blind", if you will. Their familiarity with temperature is to be traced to the successful efforts of some of their earlier scientists, who *postulated* that all ordinary objects have the properties of 'being hot' or 'being cold', properties admitting of degrees arranged on a well-ordered one-dimensional continuum, properties with the power of self-inducement and mutual cancellation in contiguous bodies, properties residing characteristically in fire, for example, for one extreme, and in ice, for example, for the other extreme. To shorten the story, they postulated what translates nicely as our common-sense conception of temperature. And this conception, they explain, has considerable explanatory power. With it they can provide uniform and coherent explanations of diverse empirical facts such as why fired kettles boil, why boiling water melts ice, why coals and fires glow, why water solidifies in winter, why people often shiver, why shivering is relieved by proximity to a fire, why snow melts in the Spring, and so on and so forth. Their theory (our common-sense) is not so powerful in this regard as is the caloric theory, not to mention the kinetic theory, but explanatory power it has. This explanatory prowess is obvious to one who enjoys no sensory access to temperature, but that scheme performs these same explanatory functions for us no less than for them.

So far then, both semantically and explanatorily, our common-sense temperature framework is as theoretical as the framework embraced by the friends of caloric. The only remaining question is whether our framework is epistemologically in a position different from theirs. It is now difficult, however, to see what the difference might be. The fact is, we are subject to a range of bodily sensations to whose occurrence we are accustomed to responding with singular judgements in the terms of the framework at issue, and by and large we get by rather well operating in this way. But exactly the same is true of the friends of caloric, the only difference being the framework used. It will not do to suggest that in our case, in contrast to theirs, our sensations are uniquely appropriate to the kinds of perceptual judgements we make, for that idea was exploded in §2. Nor will it do to seek comfort in the idea that the principles constitutive of our framework are at least true, whereas the principles constitutive of the caloric conception are false. As intimated earlier, our common-sense hot/cold theory seriously misrepresents the facts. When applied normally in certain humble cases, it encounters fatal internal difficulties, for it confidently sees only one parameter (the spectrum from hot to cold) where a more discriminating view recognizes three (*amount* of heat energy, *degree* of heat energy, and *rate of exchange* of heat energy from one body to another).

Let me try to illustrate these difficulties in specific detail. Consider the following homilies, all of which, I take it, are non-peripheral elements in our naive or common-sense conception of heat.

(1) If a given body is warmer than a second body, and that second body is warmer than a third, then the first body is warmer than the third body.

(2) If a given body is warmer than a second body, then it is not the case that the second is warmer than the first.

(3) The warmer of two bodies is the one that will cause the other to *warm up*, at least somewhat, when placed in contact with it.

(4) The warmer of two bodies (of the same weight) is the one that will warm up a third body the most.

(5) The warmer of two bodies is the one that *feels* warmer, to normal observers in normal circumstances.

These pedestrian principles are familiar enough, but the conception of heat they represent is empirically incoherent. Consider three bodies: a kilogram of green *wood* at 55 °C (130 °F), a kilogram of *iron* at 49 °C (120 °F), and a kilogram of *water* at 43 °C (110 °F). Applying criterion (3) to these objects, pairwise, we will find we must rank them, in order of decreasing warmth, as follows: wood,

iron, water. (To judge increase-in-warmth, we may use a thermometer or the palms of our hands; the empirical result will be the
same.) If we apply criterion (4), however, we shall have to rank
them as follows: water, wood, iron. This is because the temperature
difference between them is small in comparison to the wide spread
in their "heat capacities", and for all but very small third objects
it will be the water by far that shows the greatest capacity for
warming other things. At the same temperature, water will contain ten times the heat energy contained by an equivalent mass of
iron. And lastly, if we apply criterion (5) we shall have to rank
them thus: iron, water, wood. Iron's superiority as a conductor of
heat energy will make it feel warmest to the hand, and the wood
will feel the coolest, for the converse reason. Here the wide spread
in conductivity dominates the comparatively small spread in
temperature. Given principles (1) and (2), however, these diverse
results generate multiple inconsistencies. The cluster of empirical
criteria (3)–(5), while roughly univocal in most ordinary cases,
flies apart in several different directions when we examine some of
those cases closely. There just is no real property 'warmth' that
meets the collective demands the common-sense conception places
on it.

There appears to be no help for it: we are in the same position
as the "perceivers of caloric" discussed above. Our common-sense
conception here is a theory, as it happens, a false theory, a theory
that is no better off for being our current matrix of perceptual
judgement than would be the caloric theory, were it so entrenched.
The conviction that the world instantiates our ordinary observation
predicates cannot be defended by a simple appeal to the "manifest
deliverance of sense". *Whether or not the world instantiates them is
in the first instance a question of whether the theory which embeds them
is true*, and this question in turn is primarily a matter of the relative
power and adequacy of the theory as a means of rendering the
world intelligible. This pivotal lesson is urged on us now not in
some carefully contrived alien example, but right here at home, in
the case of some of the very simplest and most familiar of our
observation predicates.

The model of perception dramatized in the fable of the "perceivers of caloric" has proved encouragingly accurate so far as our
observation vocabulary for common-sense temperature is concerned. But can it serve as a model for our perceptual judgements

generally? Yes. As we move to the other perceptual modalities the sensations are different, as are the predicates and the semantic frameworks which embed them, but the epistemological situation is the same. The objective intentionalities of the various sensations peculiar to the other modalities is again a contingent matter, a relational matter, a matter of what features of the environment prompt their occurrence. And their subjective intentionalities are equally relational and contingent, being a matter of which of the many possible conceptual frameworks has been acquired as the habitual matrix of conceptual response to their occurrence. There is nothing written in the nature of things to guarantee *ab initio* the propriety of the concepts we apply. Nor can we expect that the conceptual frameworks involved with the other modalities will differ in any relevant way from the framework just indicted. Though the principles and assumptions which constitute them are indeed basic to our current conception of the observable world, their status as such affords them no proof against modification or rejection in the light of new information and fresh understanding. However familiar, entrenched, or successful they may be, they remain essentially speculative. Even at the perceptual level, therefore, our conception of the world may be myopic, confused, or even just plain wrong.

If we decide to swallow this idea, let us make no mistake about what we are swallowing. At stake here is much more than the commonplace that 'there is more to reality than meets the eye'. What is being suggested is the much stronger idea that we may be systematically *mis*perceiving reality in the first place. Rather than pursue a series of piecemeal attacks on our current conceptions, however, let us approach the whole idea by way of examining some genuine alternatives. Let us gain a taste of how *better* we might apprehend the world, perceptually. The matter of evaluative comparisons will be better pursued afterwards.

4. *The expansion of perceptual consciousness*

What we should look to here is the generally unappreciated wealth that our own sensations embody so far as their objective intentionalities are concerned. Our visual and tactile sensations of_s the presence and spatial disposition of local material objects are no less sensations of_o the presence and spatial disposition of local molecular aggregates, and it is generally obvious, perceptually, whether their constituent molecules are strongly (positionally) associated, only weakly associated, or entirely dissociated. The normal sensations of, warmth and coldness upon contact with such aggregates are reasonably reliable indicators of the mean kinetic energy (KE) of their constituent molecules.[1] For example, one is quite capable of feeling that the mean kinetic energy of the atmospheric molecules in this room is roughly 6.2×10^{-21} kg m^2/s^2, and our sensations are reliable in this regard within the range 4.8 to about 7.5×10^{-21} kg m^2/s^2. Those same sensations, in non-contact cases, are fair indicators of the energy flux of electromagnetic (EM) waves in the infrared range incident upon the skin where the sensation is located.

Let me pause here briefly to emphasize the point of these proceedings. The point is that the sensations mentioned are of$_o$ (objectively of) the novel parameters cited in exactly the same sense in which they are of$_o$ the more familiar parameters of common sense. As potential objects of observation, therefore, those novel parameters await nothing more than the same conceptual attention we currently squander on the primitive and chimerical categories of common sense.

To continue: our auditory sensations are remarkably sensitive and faithful indicators of the occurrence and properties of compression wave trains in the atmosphere – most obviously of both their wavelength (from 15 m to 15 mm) and their frequency (from 20 to 20 000 cycles per second), and derivatively, of the oscillatory frequency of whatever body or process produced them. Our current modes of thought place no useful premium on any but the most crude judgements of absolute as opposed to relative "pitch", but we can learn to do very much better than we do. (The unusual person with so-called "perfect pitch" is not a physiological freak, but an attentive and well-practised observer.) And there is much more about the world to be heard here, by way of Doppler shifts in frequency, reflection effects, refraction and interference effects, and the shape of the wave train itself. We cannot hope to match the spectacular auditory talents of the bat and the dolphin, but we can emulate them to the extent our own perceptual resources permit,

[1] In the case of associative, as opposed to dissociative (gaseous) aggregates, this should strictly speaking be mean *maximum* KE, for in such cases the energy of each particle oscillates between a kinetic and a potential mode.

and we can do substantially better than they can in many respects, for the physics of sound provides us with a conception of auditory phenomena that reveals their intimate connections with the dynamical and microphysical details of the rest of reality.

And let us not forget our olfactory and gustatory sensations, for our nose and mouth comprise an analytical chemical laboratory of admirable scope and sophistication. These provide the sensational resources for recognizing a wide variety of compounds, molecular structures/shapes/sizes, and chemical situations. Were our enthusiasm for the analytic, the descriptive, and the explanatory to equal our enthusaism for the narrowly aesthetic, these organs would render manifest to us much more of the immediate environment's constitution than they do.

And return once more to our eyes. While these are not the flawless spectrometers one could wish for, our visual sensations are eminently usable indicators of the dominant wavelength (and/or frequency) of incoming electromagnetic radiation in the range $0.38-0.72 \times 10^{-6}$ m, and of the reflective, absorptive, and radiative properties of the molecular aggregates from which it comes. And, as with sound, there is additional information tucked away here in the form of (visible) refractive, diffractive, and scattering effects. Additionally, the "incandescent" versions of certain of our "colour" sensations are fair indicators of any molecular aggregate's kinetic temperature, that is, of the mean kinetic energy again of its constituent molecules. Here, incidentally, we meet in ourselves much the same beings as those with infrared eyes discussed in §2, though our range of visual sensitivity to temperature is located much higher on the scale. In other respects, however, our own visual equipment for kinetic temperature is *superior* to theirs. For we can see at a glance the rough kinetic temperature of a star, say, independently of its apparent brightness, but they cannot, for brightness is their *only* cue for temperature. We are subject to certain "illusions" here: *spectral* emission is not always visually distinguishable from simple thermal ("black-body") emission, but collateral information will often split this ambiguity and that dual sensitivity can be as useful as it is occasionally troublesome.

These brief points barely introduce the topic of our capacity for enhanced perception, but I am here concerned only to sketch some of the features and parameters that a different culture might regard as "basic" or "maximally simple" observables. More complex theoretical features and situations must also be counted as perceptually recognizable, as we shall see. Furthermore, a word is in order here about collateral information and normal perceptual conditions. A failure of so-called normal conditions need not always frustrate our perceptual capabilities. On the contrary, non-standard conditions can often enhance them or even provide us with new ones. Kinetic temperature, for example, is much better seen in the pitch black than in broad daylight, and under various kinds of monochromatic light many features of and differences between objects become apparent that would be impossible to discern in so-called ordinary light. As often as not, different circumstances merely present different possibilities of exploitation. The trick, plainly, is to keep systematic track of them. Nor should we find disturbing the idea that collateral information about certain aspects of one's environment should modify or refine one's capacity for perceptual (as opposed to merely inferential) judgement. After all, the belief that conditions are "normal", a belief underpinning "normal" perception, is itself just a piece of collateral information, but we do not suppose that all "normal" perceptual judgements are for that reason inferential.

Altogether then, our familiar sensations simply teem with objective intentionalities over and above (or perhaps instead of) the familiar set commonly ascribed to them. Accordingly, if there is any truth to the idea that a perceiver is a being who is – by way of his acquired dispositions to make sensation-prompted singular judgements – exploiting the objective intentionalities of his sensations, then it is clear that our current modes of conceptual exploitation of sensory information barely scratch the surface, and that suitable training and education could result in different and more powerful modes of exploitation, and even in the displacement of their more primitive precursors.

Let us explore then an imaginary culture or society in which the bulk of our ordinary empirical concepts are neither used nor even remembered, a society whose "ordinary" "common-sense" conception of reality is the conception embodied in modern physical theory. In the process of language learning their children are taught to respond, in observational situations, with the relevant expressions from that theory. Where (roughly) we learn 'is red', they learn 'selectively reflects EM waves at 0.63×10^{-6} m'; where (roughly) we learn 'loud noise', they learn 'large amplitude atmospheric compression waves'; where (roughly) we learn 'is warm', they learn 'has a mean molecular KE of about 6.5×10^{-21} kg m²/s²'; where (roughly) we learn 'is sour', they learn 'has a high relative concentration of hydrogen ions'; and so on. Initially, of course, these clumsy expressions are grasped only as single units, to be parsed with profit only later, as we learned to parse 'grandma', 'television', 'Prime Minister', and the like. But their children are also taught, concurrently, the elementary inferential uses of these expressions: what they imply, and what implies them. What begins as a set of stimulus–response patterns gets progressively articulated and refined as the children acquire from their elders an appreciation of the rich fabric of entailment relations characteristic of the theoretical framework as a whole. As this appreciation grows, so does their ability to *capitalize* on those otherwise pointless patterns of linguistic response implanted initially. Their mature observation judgements, in the event, are bursting with systematic implications, both immediate and contextual, with respect to which our own observation judgements are wholly mute.

It is important for us to try to appreciate, if only dimly, the extent of the perceptual transformation here envisaged. These people do not sit on the beach and listen to the steady roar of the pounding surf. They sit on the beach and listen to the aperiodic atmospheric compression waves produced as the coherent energy of the ocean waves is audibly redistributed in the chaotic turbulence of the shallows. (Some transparent consequences of this situation are, for example, that the H_2O molecules in the shallows must have a slightly higher mean KE than those farther out, and must also contain a higher concentration of atmospheric molecules in solution.) They do not observe the western sky redden as the Sun sets. They observe the wavelength distribution of incoming solar radiation shift towards the longer wavelengths (about 0.7×10^{-6} m) as the shorter are increasingly scattered away from the lengthening atmospheric path they must take as terrestrial rotation turns us slowly away from their source. (A transparent consequence of this situation is that the wavelength distribution will swing visibly way over to the other extreme after our line of sight to the Sun is cut off and the only visible radiation to reach us is at those shorter wavelengths (about 0.45×10^{-6} m) scattered around the curve of the Earth's surface. That is to say, the western sky will be 'bluest' some time after sundown.) They do not feel common objects

grow cooler with the onset of darkness, nor observe the dew form-
ing on every surface. They feel the molecular KE of common
aggregates dwindle with the now uncompensated radiation of
their energy starwards, and they observe the accretion of re-
associated atmospheric H_2O molecules as their KE is lost to the
now more quiescent aggregates with which they collide. (A transparent consequence here is that the greatest accretion will be seen
at the surface of the most efficient radiators.) They do not warm
themselves next the fire and gaze at the flickering flames. They
absorb some EM energy in the 10^{-5} m range emitted by the highly
exothermic oxidation reaction, and observe the turbulences in the
thermally incandescent river of molecules forced upwards by the
denser atmosphere surrounding.

These observational descriptions, so arcane to us, are in no way
arcane to the people under discussion. This is the only idiom they
know. All magnitudes are conceived in rationalized units (metre,
kilogram, second, coulomb). All common substances are conceived
under their chemical and/or structural descriptions. All transformations are conceived as entries in a balanced energy ledger.
For these people the world is seen (felt, heard, etc.) to cohere, even
in its prosaic details, in ways to which our ordinary conceptions are
utterly blind.

There is an obvious difficulty here in trying to portray the im-
mediacy with which the kinds of information just surveyed can be
"read off" the world once the more powerful framework is firmly
implanted, and that is the want of any intimate familiarity on the
part of most of us with the novel ontologies and the nomic complexities of those theories which constitute it. But there is a way,
I think, in which the relevant point can be brought home, for there
is a simple theory with which almost everyone is sufficiently familiar, but which has yet to be put into *observational* gear by all but
the most devoted observers of the heavens. I have in mind here
Copernicus' theory of the arrangement and motions of the solar
system. Our minds, perhaps, have been freed from the tyranny of
a flat immobile Earth, but our *eyes* remain in bondage. Most of us
could pen quite successfully the relevant system of coplanar circles
and indicate the proper directions of revolution and rotation, but
when actually confronted with the night sky most of us have only
the vaguest idea of how to relate what we have drawn to what we
can see. And yet the structure of our system and the behaviour of its

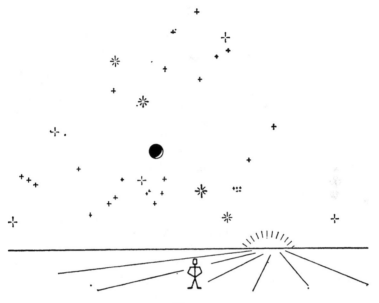

Fig. 1

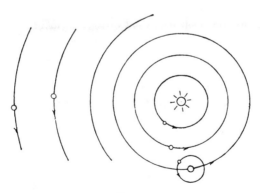

Fig. 2

elements can readily be made visually transparent, and the magnitude of the "gestalt shift" involved is rather striking. In addition to a familiarity with the Copernican view, two elements are sufficient to effect the shift. First, one must learn to recognize the several solar planets by sight. This is a two minute job, for they stand out in the night sky like beacons, and the differences between them in colour and relative brightness make them easily identifiable. And second, one must learn to reconceive all positions and motions in a novel (but natural) coordinate system for visual space, a coordinate system in which the permanent 'horizontal floor' of visual space is defined not by the local horizon, but by the *plane of the ecliptic*, the more or less common plane in which the solar planets and the Moon revolve. Let me illustrate with some sketches. Fig. 1 represents a kind of configuration not uncommon to the western sky shortly after sundown in the northern hemisphere. Aside from the Moon, the four brightest objects in sequence away from the horizon are Mercury, Venus, Jupiter, and Saturn. As can be seen, all five objects lie roughly along the same line, a line which also intersects the recently set Sun. Now the average person, though a convinced Copernican, will find this arrangement little more than curious, if it is noticed at all. But that roughly linear arrangement will become the dominant feature of the scene if it is viewed from a slightly different perspective. In order to see the situation "as it really is" (as per fig. 2), what the observer in fig. 1 must do is *tilt his head* to the right so that the relevant line (the ecliptic, in fact) becomes a horizontal in his visual field. This will help him fix his bearings within the frame of reference or coordinate system whose horizontal plane is the plane of the ecliptic, whose origin or centre point is at the Sun, and in which all of the stars are reassuringly motionless. If this can be achieved – it requires a non-trivial effort – then the observer need only exploit his familiarity with Copernican astronomy to perceive his situation as it is represented in fig. 3. Fig. 3 is of course just fig. 1, rotated through 45 °, with the basic information that a Copernican understanding supplies *drawn* in.[2] Alternatively, the situation in fig. 3 is just the situation in fig. 2 viewed edgewise, from the northern surface of the sphere riding the third orbit out from the Sun.

From the perspective of the man in fig. 3, it is plain that the axis

[2] To emphasize the desired visual point, I have allowed myself a certain graphic licence.

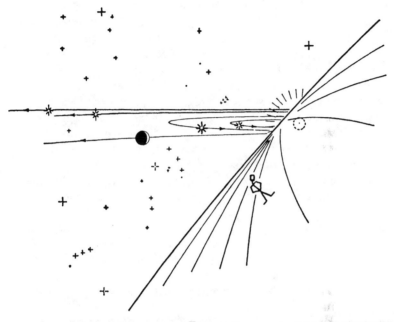

Fig. 3

of the Earth's rotation is to his right, very roughly parallel to the vertical axis of his tilted head. Keeping the stars as his fixed frame of reference, he can observe the rotation of his planet as it carries him around and away from this agreeable planetary array. In 24 h he will see the same situation again, modified somewhat by the intervening motion of everything but the Sun and the stars. What those motions will be is also plain. The position of his own planet will be translated to the right a visible amount (the position of the Sun against the stars will be shifted about 1°, or two solar diameters, to the left). The fast-moving Moon will be shifted substantially (13°) to his left, its ever-sunwards crescent swelling appropriately. And on successive nights the planets can be seen to inch along their expected paths, their reflected brilliance waxing and waning in testament to their changing distances from the Earth. Our observer can see at a glance that the Sun is much farther away than the Moon, that Venus is about to pass us " on the inside ", that the month is April, that the North Pole is up to the right, and so forth. Save for fine-grained judgements and far-future projections, his need for an ephemeris is gone. His brow need no longer furrow at the changing appearance of one entire hemisphere of his visual environment: the shifting configurations of the solar family are now visually recognizable by him for what they are. He is at home in his solar system for the first time.

I urge the reader not to judge the matter from my own spare sketches. Judge it in the flesh some suitably planeted twilight. A vertiginous feeling will signal success.

The value of this example is that the theory involved is familiar to everyone. And yet a substantial transformation in the processing of visual information is achieved once one learns to see the world as that theory bids us think of it. The items, events, and processes of the world are perceived as manifestations of a new relational order, and the continuing details of that order become perceptually obvious. It is this same kind of transformation, writ large, that I am suggesting for our habits of perception generally. The only important difference between the present example and the more radical transformation discussed earlier is that the latter is achieved by an *ontological* displacement of rather jarring proportions. The relata of the new relational order are for the most part not the items and properties countenanced by common sense. For the beneficiaries of the novel upbringing proposed, the perceptual world divides into new similarity classes, classes which cut through and across the "natural kinds" of our own observational taxonomy.[3] This point wants especial emphasis, but it is difficult for us to appreciate immediately. In reflecting on "what it would be like", the natural temptation is just to imagine a set of theoretical analogues for our current observational notions. Unfortunately, this strategy of imagination is importantly self-defeating in that the observational taxonomy thus imagined will have the same *extensional* structure as the common-sense taxonomy with which one began. But it is precisely the prospect of escaping that structure for a better one that is half the appeal of the transformation proposed. The proper way to proceed, I suggest, is to ignore the old taxonomy, conceive our various sense organs as instruments of measurement and detection, examine their multifarious capacities in that regard from within the scientific framework, and then outline as observational a suitable subset of the scientific vocabulary from there. This will prove a more complex and subtle undertaking, but such a strategy will not bind our imagination from the outset to the very taxonomy we may hope to transcend.

The basic thrust of this section is very simple. If our perceptual judgements must be laden with theory in any case, then why not have them be laden with the best theory available? Why not exchange the Neolithic legacy now in use for the conception of reality embodied in modern-era science? Intriguingly, it appears that this novel conceptual economy could be run directly on the largely unappreciated resources of our own sensory system as constituted here and now. And if my crude attempts at illustrative examples here are representative of what we can expect, the resulting expansion of our perceptual consciousness would be profound. Should we ever succeed in making the shift, we shall be properly at home in our physical *universe* for the very first time.

Do I really mean all this seriously? Absolutely. It is perhaps unfortunate, dialectically, to have to describe a startlingly maximal conceptual shift in order to get several important points across. To make the points about competing theories, ontological change, and the room for major advancement in our processing of sensory information, I needed a high-contrast example where the magnitude

---

[3] For a very simple case of this kind, and an outline of its significance, see the discussion in §3 of P. M. Churchland, 'Two Grades of Evidential Bias', *Philosophy of Science*, vol. 42, no. 3 (1975).

and nature of the conceptual jump is too large and too apparent for anyone to miss. But I will not pretend that the large-scale shift envisaged above is sociologically feasible for us in the short term, nor do I advocate that we all set about trying to teach our infants to babble in EM-ese and CM-ese. What I do advocate at the social level is that we do what we can to assist the concepts used in common sense to *evolve towards* whatever counterparts they may have in the wider conception that science has provided. (I am therefore sympathetic towards Sellars' view that "correspondence rules" connecting expressions in distinct theoretical vocabularies are best seen as proposals for the redefinition of the expressions of the less adequate theory in terms of the expressions of its more adequate and comprehensive successor.) Over the course of centuries then, we can come incrementally to complete a shift that could not have been made in one bound. Presumably this is precisely what common sense has always been doing – evolving in pursuit of the ever-advancing front of new and successful theory – so I am advocating only that we assist a process that has been underway for many millennia.

I insist then on embracing a sober position on the social side. On the individual side, however, my advocacy is less sober. It is in fact a highly instructive and entertaining diversion to try to perceive the "theoretical" in the "manifest", to try to make it emerge like a chameleon suddenly perceived against the background. There is much beauty and endless intrigue to reward the determined observer. It can also be useful, especially if one plays at reconceiving one's practical reasonings in the same theoretical terms newly applied in perception. The payoffs here, minor ones at least, prove much more plentiful than one would have expected initially. And lastly, there is the satisfaction of apprehending reality, perceptually, in ways that reflect more deeply and accurately the structure and content of that reality, of coming closer to the ideal of seeing it as it "really is".

5. *The argument from measuring instruments*

So far we have discussed a variety of possible conceptual framework/sensory equipment pairs working in tandem. We began with the conceptual economy of our ordinary hot/cold framework being fuelled quite efficiently by visual sensations (differently caused).

This was followed by the case of the caloric framework thriving on a sensational diet that normally fuels our ordinary hot/cold framework, and here the sensations were *normally* caused. This gave us a case of an overtly theoretical framework functioning as an observation framework for humans as we are now constituted, sensorily. Finding no success in our subsequent attempts to show a difference between the nature, status, or mode of use of that framework, and the nature, status, or use of our current hot/cold framework, we concluded that at least in the case of this latter framework we are indeed perceiving the world within the matrix of a theory (a conclusion reinforced by the discovery that it is a false theory into the bargain). It was then suggested that this case is typical, that the same is true of perception in the case of the other modalities as well, and this suggestion was supported by some general considerations devolving from §2. Boiled down to essentials, they come to this: if the meaning of our common observation terms is determined not by sensations, but by the network of common beliefs, assumptions, and principles in which they figure, then, barring some (surely insupportable) story about the incorrigibility of such beliefs, assumptions, and principles, our common observation terms are semantically embedded in typical fashion within a framework of sentences that has all the essential properties of a theory.

If normal perceptual judgements are instances of an established pattern of theoretical response to sensory stimulation, then the question of the propriety of anyone's perceptual judgements can be seen to turn ultimately on the question of the virtues of the theory in whose terms the responses are made. A perceptual desideratum, therefore, is that one's perceptual judgements be made, if possible, within the terms of the best available world-theory. An examination of the capacities of our own sensory system indicated that, with respect to modern physical theory, this is indeed a live possibility, and that the results of such a conceptual transformation would be highly interesting.

Though the route of our arrival is somewhat different, the view of perception here embraced is the same, in essentials, as that espoused by a number of other writers. Paul Feyerabend has argued at length against the view that theoretical terms acquire meaning only by way of receiving an "interpretation" in terms of observables. The true situation, he argues, is almost precisely the reverse.

Observation sentences – our verbal responses to sensory stimulation – would be quite meaningless save for the "interpretation"
they receive by virtue of being semantically embedded in some
conceptual framework, some *theory*.[4] N. R. Hanson has detailed
with care the perceptual changes that attend changes in background
theory.[5] In the same tradition, Mary Hesse has argued that no
predicates can function as elements in a language by dint of their
sensory associations alone.[6] They must be connected, by general
assumptions involving them, to other predicates in the fashion of
a network that is globally subject to empirical evaluation. And
Wilfrid Sellars, who holds similar views on semantic matters, has
argued in 'Empiricism and the Philosophy of Mind' that the
warrant or propriety of a conceptual response to a sensory stimulation consists (roughly) in its being an instance of a pattern or habit
of conceptual response whose inculcation and promotion it is
rational to sanction and support. Further, he makes it clear that
what makes a specific habit supportable is that it renders its subject,
like a thermometer, a reliable empirical indicator of the relevant
aspect of his environment. This analogy with measuring instruments has also been pressed, very forcefully, by Feyerabend in
'Explanation, Reduction, and Empiricism'. The obvious fact that
the states of a measuring instrument (needle positions on a dial, for
example) are of themselves mute, short of an interpretation funded
by some theory, is appealed to in illustration of the theoretical
relativity of perception and its perennial vulnerability to the
challenge of novel theories and new information.

The general line of argument – we might call it 'the argument
from measuring instruments' – bears rehearsing. We need first of
all to understand the notion of an *interpretation function*. An
interpretation function maps the distinct states of a measuring
instrument (the needle positions on an ammeter, say) onto distinct
propositions (such as, 'There is a 5 ampere (A) current flowing
in the circuit', and so on). When we calibrate a measuring instrument we determine the interpretation function we are going to

[4] See especially 'An Attempt at a Realistic Interpretation of Experience', *Proceedings
of the Aristotelian Society*, vol. 58, new ser. (1958); and 'Explanation, Reduction,
and Empiricism', *Minnesota Studies in the Philosophy of Science*, vol. 3 (Minneapolis,
1962).
[5] Norwood Russell Hanson, *Patterns of Discovery* (Cambridge, 1958).
[6] Mary Hesse, 'Is There an Independent Observation Language?' in *The Nature
and Function of Scientific Theories*, ed. Colodny (Pittsburgh, 1970).

use in relation to that instrument. Characteristically, we mark graduations, numbers, units, and so forth on a dial or scale to signal the adopted interpretation function and to assist us in applying it.

What interpretation function is appropriate for a given instrument is of course an empirical question. Interpretation functions are not sent from heaven, nor are they written *a priori* in the output states of our measuring instruments. They derive from or reflect our current understanding of the world, of its constitution and behaviour, and on occasion they must be changed, sometimes radically, as our understanding of the world grows and changes. Theories bite the dust, ontologies sink into oblivion, and new significance is attached to the behaviour of old measuring instruments.

Now it plainly will not do to suggest that each of us "sits behind" his personal battery of measuring instruments (sense organs), observes their sensational outputs, and *uses* an interpretation function in formulating his perceptual judgements. For one thing, this misrepresents badly the psychological facts of normal perception. For another, it clashes with the fact that children acquire the ability to observe and describe the world in great detail *before* they acquire any significant or explicit awareness of the richness of their sensational life, or even of its existence. And third, this representation of perception would of course leave "introspective" perception itself an unanalysed primitive, a residual mystery fundamentally different from "external" perception.

But if it is plain that perception does not involve the use of an interpretation function in the explicit manner suggested, it is equally plain, *insofar as our conceptual responses to our sensations do display determinate and identifiable patterns*, that we *embody* or *model* a set of interpretation functions, functions implanted in childhood as we learned to think and talk about the world in the fashion of our elders. And unless we are willing to plead some sort of supernatural status for ourselves, it is clear that the interpretation functions we embody or model are just as properly subjects for evaluation, criticism, and possible replacement as are interpretation functions in any other context. We may be the unwitting victims of an inappropriate set of interpretation functions. It is an empirical question whether we are; it is an empirical question which is the *right* set of functions; and it is the job of science, broadly conceived, to try to tell us what they are.

I have been using the plural (interpretation function*s*) here for several reasons. First, the human is obviously a polymodal measuring instrument. Second, under different perceptual circumstances, low light, for example, different interpretation functions should and do become operative; that is, we are able to compensate for temporary changes in the objective intentionality of our sensations. And third, as we shall see presently, changing perceptual *concerns* can bring into play different interpretation functions. The relation between human sensations and human judgements is complex and subtle indeed.

This picture of ourselves as "talking measuring instruments" escapes the awkwardness encountered earlier with "inner" perception. Introspective judgements (e.g. 'I have a visual sensation of an orange circle') can here be represented as no different, epistemologically, from perceptual judgements generally. Introspective perception involves a temporary disengagement from the interpretation functions that normally govern our conceptual responses, and the engagement instead of an interpretation function that maps (what we now conceive as) sensations, etc., onto judgements *about* sensations, etc. But no special epistemic clout accrues to those judgements merely on that account. The propriety of our introspective judgements remains contingent on the adequacy of the general conception of those inner states that those judgements presuppose. We are no better off here than we are anywhere else.

(Parallel: consider an ammeter with a graduated dial marked '5 A', '10 A', and so on. Suppose it constructed so that at the flick of a switch it flips another dial into place behind the needle, a dial marked '0.01 gauss', '0.02 gauss', and so on. This second dial is so calibrated that the needle positions on the dial now *overtly* reflect the simultaneous strength of the variable magnetic field inside the instrument, the very field whose action moves the spring-loaded needle. Our ammeter is now operating in "introspective mode". But it is not significantly more reliable in this mode than it is in the other. It may be miscalibrated here as well: in a small way (the graduations may have been badly placed), or in a large way (the whole conception of magnetic field strength as measured in gauss may turn out to be a misrepresentation of reality); not to mention the more common perils of bent needles and warped dials.)

The argument of the preceding few pages is an extremely liberal paraphrase or development of Feyerabend's presentation of the argument (see 'Explanation, Reduction, and Empiricism', pp. 36–39), but I think the spirit has been preserved. The picture of ourselves it presents is identical with the picture pieced together in the earlier sections of this chapter. Stated in terms of this section, the lesson of §3 and §4 is that, *qua* measuring instruments, we humans stand rather badly in need of wholesale recalibration.

6. *Some consequences*

From a theoretical point of view, the most important consequence of the theory of perception here embraced concerns the nature of human knowledge in general. If that theory is correct, then human knowledge is without propositional or judgemental foundations. That is, there is no special subset of the set of human beliefs that is justificationally foundational for all the rest. The traditional attempts to explicate some such fundamental asymmetry in the justificatory structure of human belief are based on a fundamental mistake. They are based on a failure to appreciate the systematic or theoretical nature of all judgements, perceptual judgements included. Given that any perceptual judgement is an instance or product of a general mode of conceptual exploitation that may itself be profoundly inappropriate to reality, we must re-evaluate the role that perception plays in the evaluation of belief and the arbitration of competing world-views. For perceptual judgements cannot provide a conceptually neutral level of factual information against which competing theories can always be effectively tested.

This conclusion does not mean that we have "cut loose from the anchor to reality". To the contrary, on the view here adopted the network of belief retains systematic causal connections with reality, connections, moreover, that carry information about reality. All we have found reason to deny is that there is a level of *judgement* that is absolutely foundational.

To allay at least some of our fears, let us approach the matter in the following way. Insofar as they represent systematic discriminatory responses to the environment, our sensory states contain information about the environment. Just what that information *is*, to be sure, is not something that can be divined *a priori* by any potential perceiver. Nor can it be divined *a posteriori*, if sensory states are considered in isolation. If they are considered collectively, however, the situation changes. With the installation or inculcation of a pattern of discriminatory *conceptual* responses to those first-level states, new possibilities are opened, for conceptual responses are themselves elements in a higher order of activity upon which there are constraints such as consistency and coherence (somehow understood). The attempts at assigning information to sensory states – attempts represented by the various patterns of conceptual response adopted – are thereby rendered open to evaluation, at least in a negative sense, and different patterns of conceptual response to sensory stimulation may be tried in hopes of avoiding the difficulties encountered by the initial attempts. In some such humble fashion, we might presume, does the activity of empirical thought have its beginnings. And it has its end, we may conjecture, in maximizing the variety and sheer amount of (putative) information winnowed from the flux of sensory states in essentially the same fashion. It seeks and continues to seek information in what confronts a perceiver initially as just so much noise.

These remarks hardly constitute an adequate normative epistemology, nor will they allay all fears. But they will serve to illustrate that the denial of a foundation at the level of *judgement* need not deny us a continuing and critical contact with reality. It requires only that we reconceive what that contact consists in. This matter will be taken up again in later chapters.

Another matter, of comparable importance, arising from the themes of this chapter concerns the fate of our common-sense ontology. What status does reason demand we actually assign to the speculative conception of reality embodied in our common-sense conceptual framework, if the emerging picture of reality provided by modern-era science is accepted, provisionally, as true? This question clearly needs dividing, for the common-sense conception of reality is a loosely integrated patchwork of subtheories rather than a unified monolith, and parts of it may fare better than others in the crucible of enlightened criticism. Additionally, though the "emerging picture of reality provided by modern-era science" is downright stupefying in its power and generality, we cannot pretend that the picture is complete, unified, or unblemished in its empirical adequacy. The picture, in short, is still very much in the business of emerging. And even if we take these complexities into account, the divided issues suitably refined will require for their settlement a much clearer consensus on the matter of theory displacement versus theory reduction than philosophers have so far achieved. Though we might all embrace the principal tenet of scientific realism that excellence of theory is the measure of ontology, this principle will not suffice unaided to settle all of our ontological questions. The class of φs, say, might be postulated by what is at best a weak and superficial theory, and yet maintain its ontological credentials intact, if that theory reduces cleanly to some superior theory whose virtues compel our assent. Accordingly, the strictness of the conditions imposed on intertheoretic reduction prove of central importance here.

In sum, an answer to our question that is both easy and general is not to be expected. Just the same, the profound complacency most philosophers display concerning the status and/or the staying power of the common-sense conception of reality appears to me to be ill founded in the extreme, and I should like to close this chapter with some brief critical remarks.

The most popular position on this matter maintains that our common-sense conception is a narrower conception of reality, a less powerful conception than the one modern theory provides, but that so far as it goes, it is not wrong or false in any fundamental respects. Though it may represent reality only partially and selectively, it is not a *mis*representation of reality. More specifically, this view continues, the items and features countenanced by the common-sense theory of reality can be identified with certain of the items and features countenanced by modern theory in such a fashion that the assumptions and principles of the former become simple consequences of the assumptions and principles of the latter. That is, our common-sense ontology will survive by being *reduced* to the more penetrating ontology of modern physical theory.

On what, precisely, is this reassuring view founded? A prime element in the complacency of many is still, I think, the following conviction. 'The claim on our belief held by modern physical theory derives essentially from its success in explaining the behaviour and constitution of reality as it is conceived by common sense. And it has achieved this success by conceiving theoretical analogues for, or descriptions of, the elements of the common-sense ontology such that the novel theory that embeds them entails those features (or their theoretical analogues) of the elements of our ordinary ontology that stood in need of explanation. The conditions needed for the reduction of common sense are therefore built into the aims of the scientific enterprise from the outset, and the success of that enterprise will guarantee that they are met.'

This view is not (quite) wholly mad, but in this naive form it is transparently indefensible. We have encountered this view before in the worried concern of our calorified friends over the apparent irreducibility of the caloric theory to the corpuscular/kinetic theory, a concern born of their quaint conviction that the believability of the latter resides essentially in its ability to explain the "facts" as conceived in terms of the former. Successful theory must indeed explain the facts, but it would be madness to make it a constraint upon acceptable theory that it explain the "facts" as they may currently be conceived by us. The question of what makes the starry sphere of the heavens turn daily – a live question for two millennia – would never have been disposed of had we ruled out of court any dynamical theory that denied the motion or the existence of that sphere. The "facts", as currently conceived and observed by us, form the starting place for theoretical inquiry, but its successful pursuit may well reveal that we should vacate that starting place as hastily as possible. Large-scale intellectual progress will involve the wholesale rejection of old *explananda* as frequently as it involves the wholesale introduction of new *explanantia*.

The poverty of the view under attack can be summarized as follows. The demand it places on successful or acceptable theories – that they explain or reduce the facts as conceived within common sense (or within theories already "established") – assigns to the framework of common sense a significance beyond what it deserves. That framework, after all, is just *the theory that got there first*, and this is hardly sufficient reason to demand that all subsequent theories treat it as a touchstone for their own adequacy. To be sure, it will be counted a plus, all other things being equal, if a theory successfully reduces its apparently virtuous precursors, but this is most definitely a non-essential virtue. The reduction of our commonsense ontology is therefore not guaranteed by the nature of the scientific enterprise itself.

Beyond affirming these general points, I will not attempt to dogmatize on whether this, that, or the other area of our commonsense ontology is both displaceable by and irreducible to this or that area of the scientific ontology. There is little profit in this, I have discovered, since the conditions on genuine reduction are so vague and elusive, and since philosophers often show a decided tendency to relax their conditions on reduction to whatever extent they discover is necessary to preserve whatever area of the common-sense ontology they see threatened. This selective concern with the cross-theoretic similarities rather than the cross-theoretic differences is, in a sense, quite all right, for in such continuities and conceptual parallels as we may find holding across the conceptual divide lies our explanation of why the older scheme worked as well as it did. The mistake lies in thinking that our ability to point to such parallels and to provide such explanations amounts to a *vindication* of the older scheme, in the sense of providing a rationale for continuing to embrace it. The nature of this mistake will be discussed in more detail in §11.

3
The plasticity of understanding

7. *The analytic/synthetic distinction*

The growing recognition that the analytic/synthetic distinction is as unreflected in linguistic fact as it is recalcitrant in linguistic theory has made the discussion of meaning interesting again. The familiar picture of a sharply delineable conceptual framework distinct from and presupposed by the edifice of merely empirical belief, a framework whose girders are analytic truths and whose joints are concepts rigidly defined thereby – all this must be swept away, to be replaced by the holistic and dynamic picture of the evolving network of all of a man's beliefs, beliefs no longer differentiated by any exclusive semantic credentials or unique epistemological status. The older picture must be swept away not because there are no semantically important differences between one's beliefs – of course there are – but rather because its explication of what those differences consist in is confused, mistaken, and explanatorily sterile.

The plausibility of the analytic/synthetic distinction lies principally in the fact that for certain of the sentences we accept – for example, 'All bachelors are unmarried males' – we find it plausible to insist that they do not admit of a denial that is consistent with our current understanding of the terms they contain. And this has suggested to many that the meaning of the terms they contain is the source or ground of the truth of such sentences. From this results the familiar conception of analytic sentences as those sentences true solely in virtue of the meanings of the terms they contain, and the usual litany concerning their necessity and lack of any empirical content.

But we should not have been taken in so quickly, for none of this is supported by the evidence, if one looks at it closely. Consider first the basic intuition, felt for some sentences, that their denial would be inconsistent with one's current understanding of one or more of the terms they contain. And let us concede that there are in fact such sentences. Of itself, this fact should not move us to find the truth of such sentences somehow rooted in our understanding of the terms they contain, for there are a great many sentences meeting the condition just conceded where we feel no temptation to count them as analytic, as necessarily true, as knowable *a priori*, or as void of empirical content. For an example, consider the following set of sentences and the simple conception they embody.

(1) Phlogiston is an elemental substance.
(2) Phlogiston forms compounds with other substances.
(3) Combustion and calxification both consist in the release of phlogiston from a compound substance containing it.
(4) The release of phlogiston is induced by high temperatures.

Let us put ourselves in the position of a proto-chemist who has just formulated this theory in hopes of rendering intelligible the details of certain familiar physical transformations. Aside from a few leading assumptions as to which substances are phlogiston compounds (wood, metals), the set of assumptions (1)–(4) effectively *exhausts* our conception of phlogiston, our understanding of the term 'phlogiston'. These are the sentences that introduce the term into our general linguistic commerce, and it has no source of semantic identity, initially at least, beyond these. In such a case the denial of one or more of these assumptions would indeed be inconsistent with our understanding of the term 'phlogiston'. And yet, though we ourselves are the authors of these assumptions, we feel no temptation to claim that they are necessary truths, that they are empirically irrefutable, that they are true solely in virtue of the meanings of their terms. On the contrary, we are aware of the highly distinct possibility that one or more, or even *all* of them, are false. And rightly so, for as we know, that is how in fact the story ended.

So far then, the basic intuition cited earlier will not serve to distinguish such paradigms of analytic necessity as 'All bachelors are unmarried males' even from such paradigms of synthetic contingency as 'Phlogiston forms compounds with other substances'. A given sentence may not admit of a denial consistent with one's current understanding of its terms, but this provides no guarantee that the sentence is even *true*, let alone necessarily true, let alone true solely in virtue of meanings.

The case of 'phlogiston' is not unusual. It is often the case that sentences which express or form part of the meaning of a given term turn out to be false. Consider the even more striking example of 'Atoms are indivisible particles'. In such cases we speak of confused, mistaken, inappropriate, or false conceptions. And properly so, for our conceptions do not form an empirically uncriticizable "given" fit to serve as a *source* of truth in its own right. On the contrary, our conceptions are themselves continually subject to evaluation, criticism, and modification as the process of learning about the world proceeds. That conceptual change takes place is of course something that defenders of the analytic/synthetic distinction have always been willing to concede, but its existence poses a serious problem for them just the same. Why does it happen at all? Why do we find it advisable on occasion to modify our concepts, to change the meanings of some of the terms we use? Since the "analytic" sentences that express or embody those meanings are held to be empty of empirical content, a change of meaning cannot be forced on us by any empirical facts. And since any proposed new set of analytic sentences (expressing new meanings) will be equally innocent of any empirical content, they can have no empirical virtues to make them preferable to the set with which we began. Conceptual change must therefore occur for reasons other than the various empirical considerations that motivate changes in our synthetic beliefs. Of what sort are these supposedly special reasons?

Despite its importance, this is not a question that has been much confronted by defenders of the analytic/synthetic distinction, but to the extent that it has been confronted, the answers have been along the following lines.[1] A change in the meanings of some of the terms we use to describe the world often allows us to express a *simpler* or *smoother* overall account of it. For example, we may narrow our earlier understanding of the term 'fish', under which 'A whale is a giant fish' was analytic, so that this sentence comes to express a synthetic falsehood. The payoff is that we are then able to state simply a great many useful generalizations about fish that would have to have been counted false had we continued to count whales as fish. And with our acceptance of 'A whale is a giant

---

[1] See Rudolf Carnap, 'Empiricism, Semantics, and Ontology', in *Meaning and Necessity*, 2nd edn (Chicago, 1956), esp. pp. 206–8. Also, C. I. Lewis, 'A Pragmatic Conception of the A Priori', *Journal of Philosophy*, vol. 20 (1923); reprinted in *Meaning and Knowledge*, ed. Nagel and Brandt (New York, 1965).

aquatic mammal' in place of the sentence relinquished above, many facts about whales become simple consequences of our general beliefs about mammals. What motivates a change in our meanings then, as opposed to a mere change in our beliefs, is the greater simplicity and coherence it permits in our *description* of empirical reality.

If we swallow this story – as an account of the *contrast* between empirical and non-empirical considerations – then we have swallowed a hoax. As Quine has pointed out,[2] any sentence could be held to be true, no matter what the countervailing evidence, if one were willing to pay the price of increasing complexity and decreasing coherence in the rest of one's beliefs. Accordingly, the avoidance of such unhappy adjustments and accommodations, and the correlative pursuit of overall simplicity and coherence, is unquestionably a central consideration – perhaps the prime consideration – even in one's evaluation of avowedly synthetic sentences. The undoubted relevance of such considerations in the example discussed above will therefore hardly serve to distinguish the kind of change it involves as non-empirical. On the contrary, it makes them look very empirical indeed. As a result of learning more about fish (that they breathe water, lay eggs, etc.), and more about whales (that they breathe air, give birth to live young, etc.), we come to realize that whales are not fish at all, but rather mammals bearing a superficial resemblance to fish. To describe the matter in this way is *not* to deny that our understanding of the terms 'whale' and 'fish' may have suffered some change in the process, but it is to reject the empirical/non-empirical dichotomy some would have us swallow in explication of that change.

The poverty and epistemological irrelevance of the supposed analytic/synthetic distinction can be most pointedly illustrated, I think, in the following way. Consider two people, A and B, whose linguistic behaviour is almost identical. They share, initially at least, identical dispositions to accept observation sentences, and they assent to all the same non-observation sentences in every area of belief, save one. On the semantic and modal properties of the vast horde of sentences they hold in common there is almost complete disagreement. Somewhat remarkably, for almost every general sentence he accepts, B insists that it is necessarily true,

[2] W. V. Quine, 'Two Dogmas of Empiricism', *The Philosophical Review*, 60 (1951); reprinted in W. V. Quine, *From a Logical Point of View* (New York, 1963).

analytic, true solely in virtue of the meanings of the terms it contains. In sharp contrast, A refuses to concede, for all but a very few of those shared sentences, that they are anything more than the merest of synthetic contingencies.

Supposing both A and B to be rational, let us now consider the subsequent evolution, in each of them, of this initial set of shared sentences, as A and B share a common experience of a common world and adjustments inevitably come to be made. A, we may suppose, represents someone like ourselves, and the adjustments he makes in the set of sentences he accepts are made with an eye to maximizing the simplicity and coherence of his overall account of the world. But B is nobody's fool either. Though apparently "frozen into" the initial set of sentences in a way that A is not, B none the less values overall simplicity and coherence as highly as does A, and he is every bit as insightful as A in perceiving just what modifications in the initial set of sentences will best achieve those goals, given the particular set of problematic observation sentences to which they have just given their common assent. Accordingly, B *makes* the required adjustments in the initial set of sentences, the *same* adjustments as those made by A. *For so long as B values simplicity and coherence above retaining his initial meanings, there is no reason why his adjustments should differ in any respect from A's.* B of course describes those changes differently. As B tells it, B has changed the meanings of some of his terms; he is the subject of a minor conceptual revolution. A, on the other hand, claims only to have changed a fair number of his beliefs. The result, in sum, is that they again share assent to identical sets of sentences, having made identical adjustments for identical reasons, but they disagree still on the semantic and attendant modal properties of those sentences.

What significance then are we to attach to such disagreements? Precisely none, at least from an epistemological point of view. One might have expected B to be more stubborn about avoiding changes in meaning, and hence to have made adjustments different from those made by A. But this expectation violates the assumption that B is rational. Though he assigns analyticity ("definitional status") more freely than does A, this does not mean that he must display any resistance to changing those assignments ("definitions") if the pursuit of his larger epistemic ends requires it. If he is rational, he will have no such resistance. And so long as he is willing to change his so-called meanings, the range of adjustments open to him equals the range open to A, and the actual adjustments he makes need not (should not) differ in any way from A's. This does mean, however, that assignments of analyticity count for nothing, epistemologically, in any rational man. Though we may concede to B that in making the adjustments at issue he has changed some of his meanings, it is plain that the question of what meanings he *should* attach to his terms was and remains very much an *empirical* question, a question not obviously distinct, moreover, from the simple question of which overall set of sentences he should embrace in pursuit of his general epistemic ends.

Common intuitions concerning the analyticity of this or that sentence do reflect something, however, if not the existence of a special non-empirical substructure embedded in the network of our beliefs generally. Primarily they reflect the relative *semantic importance* (with respect to a term T) of the sentences we accept (containing T). Semantic importance is less an individual than it is a social matter. Roughly speaking, the semantically important sentences are just those sentences on whose acceptance (at time *t*) the smoothness and efficiency of one's verbal commerce in a community is maximally dependent (at time *t*).

Let me try to explain. What is it, after all, that permits two speakers to understand one another, to converse freely and efficiently with one another? Given that they share identical vocabulary, syntax, and dispositions to draw formal inferences, surely it is the systematic similarity in the sets of sentences they respectively accept, and the correlative similarity in the material inferences they are therefore disposed to draw. To the extent that two speakers are similar in this respect, to that extent can each treat the other's assertions as if they were sentences of his own, so far as the repercussions of accepting or rejecting them are concerned. But this cannot be the whole story, for in general, different speakers are disposed to assent to widely divergent sets of sentences. It would be a rare twosome indeed who did not harbour flat disagreement on a fairly substantial number of sentences, and the divergence between them will be wider still when we consider the much larger set of sentences to which one will give assent (or dissent) while the other shrugs his shoulders in disinterested ignorance. And yet they may converse with one another with untroubled fluency even so.

Clearly the possibility of smooth and efficient communication depends on the sharing of some sentences more heavily than it depends on the sharing of others. The traditional view is quick to identify these as the *analytic* sentences: analytic sentences embody meanings, and it is shared meanings that makes communication possible. So long as two speakers share the same analytic sentences, their merely synthetic beliefs can diverge widely and their ability to communicate will remain intact.

Here the semantic importance of certain sentences is explained in terms of their proposed analyticity. But this explanation is idle, and for reasons additional to the familiar problem of providing a defensible explication of what analyticity consists in. Recall A and B of several paragraphs ago. By hypothesis, they had no (or a negligible number of) analytic sentences in common. They do share assent to identical sets of general sentences, but while almost all of these were analytic in B's idiolect, they were synthetic in A's idiolect. And yet A and B will have no trouble communicating. Indeed, they might converse freely and efficiently for years before ever discovering the "difference" in the "semantic status" of their respective general beliefs. Whatever it is about their shared sentences that makes effective discourse between them possible, it is not to be explicated in terms of their "analyticity", for it matters nothing to effective discourse whether they be analytic or no. Moreover, as the phlogiston case attests, it is not even required that they be *true*.

What is it then that makes certain sentences semantically more important than others? A number of things. To begin with, the mere fact that a sentence enjoys universal acceptance within the relevant community earns for it an importance it would otherwise lack, for the network of all such sentences defines the common inference patterns each person shares with and tends to expect of every other member of the linguistic community. Among these, of course, the general sentences will tend to enjoy a greater importance than the singular sentences, by reason of their more prominent role in the business of drawing material inferences. Also, those universally shared general sentences that enjoy the most frequent invocation, tacit or explicit, in the drawing of everyday inferences must be assigned a semantic importance greater than that of their less frequently invoked fellows. Relative "isolation" is also an important factor. If a shared sentence containing a term T happens to be one of a very small number of such sentences containing T, it must be reckoned more vital than were it just one member of a large and active cluster. Greater importance will also attend a sentence if that sentence is the most commonly used formula for *introducing* the relevant term to the vocabulary of those speakers still innocent of its use.

The common thread that unifies these diverse conditions is the importance for communication of the sentences that meet them. A speaker who uses a term T in a fashion inconsistent with what happen to be (at that time and in that community) the semantically important sentences involving T, is a speaker whose efforts at communication will probably win confusion for his audience and frustration for himself. Unless he can exploit whatever understanding he does share with his audience, and can relate to them a story sufficiently compelling to win a change in their semantically important convictions involving T, he will find himself dismissed as someone who does not know the meaning of that term.

The notion of semantic importance will perform this much of the job that analyticity was designed to perform, but here the similarity ends. The semantically important sentences, as characterized above, have no claim to any special modal or epistemological status. At the social level such sentences do enjoy a non-trivial significance. Collectively, they embody the baseline of linguistic contact between speakers, the common denominators each speaker expects to find, and to be able to exploit, in other speakers. But that is all. And at the individual level the significance of such sentences is drastically reduced. At most, they will tend to be the maximally familiar of one's accepted sentences.

On the account of things here proposed, it is plain also that the semantic importance of a sentence is a matter of *degree*, and may indeed vary from subgroup to subgroup within the linguistic community at large. Equally, *change in meaning* must here be counted a matter of degree as well, a matter of the extent of the revisions made in the (anyway fuzzy) set of sentences semantically important for whatever term is at issue, and of the degree of semantic importance of the sentences actually relinquished. The continuity here asserted is directly reflected in the large number of "borderline" sentences whose existence has always plagued defenders of the analytic/synthetic distinction, the problem being the want of any effective criterion for deciding whether or not their denials would be inconsistent with or would require or amount to a change in the meanings of the terms they involve. The search for any such sharp criterion must be discounted as vain, for the modal, epistemological, and semantic dichotomy it seeks to explicate is an illusion.

Finally, it will not improve matters to seek solace in the idea that 'analyticity' is an *idealized* notion, like 'point mass', which applies only imperfectly to the fluid elements of a natural language. The motive behind such a position would presumably be the belief or the hope that said notion, like 'point mass', could none the less play an important explanatory role in our understanding of the kinematics and dynamics of our intellectual activity. But this hope is vain also. As the parallel intellectual evolution of A and B attests, an analytic/synthetic distinction will be without detectable significance for epistemology. And as the ability of A and B to converse freely further attests, an analytic/synthetic distinction will be without detectable significance even for *semantic* theory!

8. *Meaning and understanding*

It will prove useful in what follows to draw a distinction between an individual's understanding of a given term T, and the meaning of T in the language of his community. The former can be represented, in first approximation, by the set of sentences containing T accepted by the individual in question. The latter is represented by the set of sentences semantically important for T in his linguistic community. When the latter forms a subset of the former, in a given individual, that individual will be an effective and non-deviant speaker of that community's language so far as his command of T is concerned.

Both notions need refining, however, if certain obvious complexities are to be handled. In evaluating the semantic importance, for a term T, of this or that sentence, it is not strictly speaking the entire community that should concern us. For example, the linguistic dispositions of children who are still in the process of learning the language will not, for obvious reasons, be considered as among the determinants of semantic importance. And with respect to some terms, the "preferred" group whose linguistic dispositions collectively determine semantic importance will be quite small indeed. Consider 'muon', 'nucleotide', 'dyslexia', 'torus', 'photosphere', and even 'elm'. For some of these terms, most English speakers have no understanding of them whatever, and for the others, only the most truncated grasp. The average speaker's understanding of 'elm', for example, more-or-less begins and ends with 'An elm is a kind of tree'. And yet all of these are terms in the English language. In such cases we defer on questions of meaning to the "experts", to those who profess to know about muons, nucleotides, and elms.[3] That is, the group within which semantic importance is to be estimated is limited to those individuals who at least *have* some non-trivial understanding of the relevant term, whatever that understanding may consist in. The lexicographer may properly leave out of account, however, the odd plainly deviant speaker (e.g. the man for whom muons are citizens of Mu, a prehistoric community of advanced achievements whose island sank beneath the sea after a war with the Venerians, etc., etc.).

Though we have allowed the size of the relevant constituency to vary considerably, the notion of *acceptance* remains central to the account of semantic importance here proposed. That notion needs qualification in one further dimension, however. Theoretical terms that have died and those just in the process of acquiring life provide the motivating examples. Consider our earlier example of 'phlogiston'. The point of that example, and there are many others, was that the sentences semantically important for that term are false, and known to be false. Not surprisingly then, they are no longer accepted by anyone. But the term is still meaningful; indeed it means now exactly what it meant before. And that is because there are no sentences currently important for that term beyond those that *were* semantically important for it in its heyday. The term "died": the network of sentences collectively sustaining it ceased to evolve and was simply abandoned. It therefore has no semantic identity beyond what that network supplied. We concede it a specific meaning then, on the strength of its specific past. In sum, past acceptance can be sufficient for present semantic importance, so long as there is no current competition for control of the term involved.

The case of new theoretical terms, or rather, of the basic sen-

[3] Cf. Hilary Putnam, 'The Meaning of "Meaning"', *Minnesota Studies in the Philosophy of Science*, vol. 7, ed. Gunderson (Minneapolis, 1975). Reprinted in Hilary Putnam, *Mind, Language and Perception* (Cambridge, 1975).

tences of the novel theory that embeds them, requires of us a related qualification. For the sense in which the postulates of a new theory are accepted is an attenuated one. At best, they are "assumed" for the purposes of testing and evaluation. Semantic importance remains delineable, however. Confining ourselves to the subgroup who use the terms at issue, the semantically important sentences will be those most widely assumed in the attenuated sense conceded, those most commonly presented in introducing the theory to others, etc. These are the sentences that *will* enjoy wide acceptance if the theory proves successful. At the risk of straining a parallel, we might say that the prospect of future acceptance can be sufficient for present semantic importance, if control of the relevant term is not preempted by current competition.

So far for semantic importance and for the notion of meaning erected on it. Shift attention now to the notion of a particular individual's *understanding* of a term T. The first approximation given above would have us represent a given individual's understanding of T by the set of sentences he accepts containing T (hereafter, his T-set). However, a more useful notion will result if we acknowledge certain differences among the sentences in a given speaker's T-set, the point being to liberalize somewhat the criterion for *sameness* of understanding across different speakers.

Some of the sentences in any speaker's T-set will be more important than others, in the sense that their denials would require more extensive readjustments and revisions in the body of his beliefs generally, or would involve a greater loss of explanatory power, than would the denials of others. These are the more deeply "entrenched" sentences, or, as we shall call them here, the T-sentences that are *systemically important* for that individual. This notion of systemic importance parallels in certain respects the notion of semantic importance, but it is crucial that we observe the very substantial distinction between them. The semantic importance of a sentence is exclusively a *social* rather than an individual matter, and its relevance is confined to the theory of communication. By contrast, the systemic importance of a sentence is primarily an individual matter, and its relevance is overtly epistemological. To be sure, many sentences will be simultaneous examples of both notions, but other sentences will rate high in one dimension and very low in the other. In illustration of this, consider the plot in fig. 4 of the relative positions of various sentences. The semantic

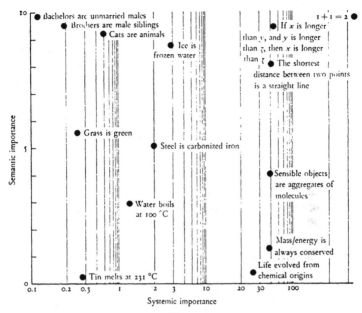

Fig. 4

importance represented by the vertical axis is to be taken as esti-mated for the class of English speakers generally. The systemic importance represented by the horizontal axis expresses the relative entrenchment, of the sentences displayed, in the belief network of this writer, but I presume the results will not differ unintelligibly from how most of my readers would place the same sentences.

Judging from the graph, it would appear that semantic and systemic importance are substantially independent variables, for sentences are scattered throughout all positions. This is because the horizontal axis plots systemic importance relative to *all* sentences. 'Batchelors are unmarried males', for example, rates very poorly on this particular scale. However, if we were to construct another graph whose horizontal axis plotted systemic importance *relative to other sentences containing 'bachelor'*, the sentence cited would rate considerably higher. That is to say, for any given term T, there does appear to be a significant positive correlation between the semantic importance of a sentence containing T, and its systemic importance (for most speakers) relative to other sentences containing

T. This is not particularly surprising, of course, for the former will tend to be the social distillation of the latter.

The details of fig. 4 must not be taken too seriously. Both the horizontal and the vertical plottings are very sensitive to variations in the persons and groups for which they are estimated, and the metric imposed is quite arbitrary. Additionally, the estimates presented are merely ballpark stabs on my part. They can be trusted to reflect, even on assumptions kind to me, only gross differences in relative position.

Even the gross results are interesting, however, for the contrasts they display, and it will not have escaped the reader's attention that there is a decided tendency for classically "analytic" sentences to be clustered towards the extreme upper left, and for familiar examples of "synthetic *a priori*" sentences to surface at the extreme upper right. It is tempting to see in these distributions the materials for an explanation of the variety of intuitions philosophers have felt when they set themselves to dividing sentences according to the old taxonomies of the analytic versus the synthetic, and the *a priori* versus the *a posteriori*. Consider, for example, the venerable criterion used to isolate *a priori* sentences: the inconceivability of their denials. Clearly, the ease with which a semantically non-deviant story can be imagined – in which a sufficient motive for denial arises naturally from the storied conditons – will be at a minimum for general sentences in the upper left quadrant. Ease of imagination will rise quickly as we leave that area, but will start to shrink again as we move farther and farther to the extreme right. For the material interconnections (with the rest of one's beliefs) on which plausible falsifying stories depend are at a minimum for sentences at the extreme left; and to the extreme right they are there in abundance but so deeply entrenched that denial short of a major revolution becomes impossible, and unusual creativity and theoretical skill will be required even to conceive of a coherent alternative. A further brake on one's enthusiasm for denial is provided by the high semantic importance that "synthetic *a priori*" sentences share with the "analytic" sentences. The only difference lies in the much greater systemic importance of the former. The appearance of empirical vacuity, the hallmark of analyticity, will therefore be lower for sentences towards the right. Hence the temptation to cast them as synthetic.

I suggest then the following explication of that class of sentences philosophers have tried to understand in terms of the notion of analyticity. They are simply those sentences that are simultaneously very high in semantic importance and very low in systemic importance. In my own case, this two-dimensional explication was originally spawned by a dissatisfaction with Quine's offhand explication (in 'Two Dogmas') in terms of relative entrenchment solely. (The sentence 'All bachelors are unmarried males', though a paradigm of "analyticity", is an implausible example of deep theoretical entrenchment.) The discovery, upon plotting familiar sentences, that the "synthetic *a priori*" might also be captured as those sentences very high in both systemic and semantic importance was an unexpected bonus, and gave added credence to the rough account proposed. But let us now leave the matter of analyticity behind. So far as explicating that particular notion is concerned, the account just outlined differs only marginally from the familiar account proposed by Hilary Putnam,[4] and its further pursuit here would be a digression from our main task.

Consider now, for a given term T, the systemically more important sentences in a given individual's T-set. These need not, we should note, possess much systemic importance on any absolute scale. Terms for non-natural kinds, for example, tend to have relatively little systematic significance; their T-sets tend to be exhausted by sentences of relatively low systemic importance. Still, even for those T-sets the systemic importance of some elements will exceed that of others (general sentences over singular sentences, for example), and it is to such subsets that we should now direct our attention.

Let us (re)represent an individual's understanding of a term T by the systemically more important members of his T-set. Accordingly, we shall say of two speakers that they have the *same* understanding of T just in case the "preferred" subsets of their respective T-sets have the same membership. This will allow those same two speakers to differ substantially in the systemically less important elements in their respective T-sets (as of course most of us will, the mass of idiosyncratic trivia being considerable in all of us), while still holding a common understanding of T.

This refined notion of understanding, and the liberalized identity criterion it permits, is clearly an improvement over its precursor.

---

[4] Cf. 'The Analytic and the Synthetic', *Minnesota Studies in the Philosophy of Science*, vol. 3, ed. Feigl and Maxwell (Minneapolis, 1962), esp. p. 365. Reprinted in Putnam, *Mind*.

But we must not misestimate the nature of the gain. Identity of understanding has here become a pragmatic matter, a matter of how far down (in the direction of decreasing systemic importance) in our respective T-sets we are willing to dig before we cease to care about whatever differences further digging might unearth. For some issues then, we may count our respective understandings of T as identical, while for other issues we shall count them as non-identical, some small difference between us looming large because it happens to bear on the specific issue at hand.

Let us review where we stand. We have before us a notion of meaning, and of sameness of meaning. A term T has the same meaning for group α as for group β just in case the respective sets of sentences that are semantically important for T in each group have the same membership. Also, we have before us a notion of individual understanding, and of sameness of understanding. A person x's understanding of T is the same as person y's just in case the systemically more important elements of their respective T-sets are identical.

Supposing that the minor difficulties confronting them so far have been surmounted acceptably, we must confront a respect in which both notions are fundamentally inadequate. The inadequacy is identical in both cases, and it consists in the fact that my characterization of each is simultaneously too strong and too weak. Consider the problem first as it confronts the characterization to this point of sameness of meaning across distinct groups. It is plain that meaning can be the same and yet the relevant sets of sentences semantically important in each group be different. Suppose, for a given term T, that the relevant set in group α differs from the relevant set in group β by just one sentence: the former contains a sentence S_α where the latter contains a sentence S_β. S_α and S_β differ in just one term: S_α contains a term T_α where S_β contains a term T_β. Now if T_α and T_β happen themselves to be synonymous, there need be no problem about the original term T having the same meaning for both groups, but the notion of meaning proposed above will preclude this, since the sentences semantically important for T in each group are not strictly speaking identical. More generally, consider 'baum' in German and 'tree' in English. These have the same meaning, but here the relevant sets contain *no* sentences in common.

It is plain also that the relevant sets can be identical, and yet the meaning of T be different in each group. Suppose, for example, that although the relevant sets of T-sentences in each group are identical, one of the terms in the α-set has a meaning quite different from its homonymous counterpart in the β-set. Here the meaning of the original term T will differ for each group, despite the identity of the sentences semantically important for T in each. Accordingly, the requirement which our characterization imposes is too weak as well. (Note that these same difficulties would also arise if we were working with the notion of "analytic" sentences in place of the notion of semantically important sentences. Our picture here is in error, but not because it has forsaken analyticity.)

Of course, questions of meaning can be and commonly are settled by reference to a notion of meaning such as the one we have found wanting, but only because those questions commonly arise in situations where the speakers already share identical or closely similar linguistic backgrounds, where a specific background language is already presupposed. The plausibility of the specific notion of meaning proposed, to the extent that it was plausible, was contingent on our considering the matter only at this parochial limit. In fact, as our counterexamples indicate, the basic unit of meaning must approximate an entire language. For the meaning of smaller units can be decisively specified only against the background provided by some such maximal unit. The semantic identity of a term derives from its specific place in the embedding network of the semantically important sentences of the language as a whole. Accordingly, if we wish to speak of sameness of meaning *across* languages, then we must learn to speak of terms occupying *analogous places* in the relevantly *similar networks* provided by the respective sets of semantically important sentences of the two languages at issue.

Precisely parallel difficulties attend our earlier characterization of sameness of understanding across individuals, and the lesson to be drawn mirrors the lesson drawn above. The basic unit of understanding must be something approximating an entire idiolect. An individual's understanding of a term can be decisively specified only against the background provided by the entire interlocking network of (systemically more important) sentences he accepts. Therefore, if we wish to speak of homodoxy (sameness of understanding) across idiolects, we must again think in terms of corresponding nodes in sufficiently parallel networks. In this way we are led to a *holistic* conception of both meaning and understanding.

Can non-metaphorical sense be made of this talk of "corresponding nodes" and "parallel networks"? Yes it can, and the basic idea is very simple. It is most clearly illustrated in the case of two languages that differ only phonetically. The similarity of their "intensional structures" consists in the fact that there is a mapping of the terms of the first onto the terms of the second such that the images, under the mapping, of the sentences semantically important for each term in the first language are exactly the sentences semantically important for its paired term in the second. Such a mapping is nothing other than a mutual translation of the two languages, in this case, a *perfect* translation.

To be sure, languages usually differ more substantially than in the case just described. Consider then the following more general notion of translation.

> (i) An *optimal translation* of L_α into L_β is a mapping of the terms and syntactic forms of L_α into the terms and syntactic forms of L_β such that (1) the image, under that mapping, of each grammatical sentence of L_α is a grammatical sentence of L_β, and (2) the images, under that mapping, of the sentences semantically important for each term T_α in L_α are exactly the sentences semantically important for its paired term T_β in L_β.

With this notion in the background, a suitable definition for synonymy across languages is readily framed.

> (ii) Two terms, T_α in L_α and T_β in L_β, are *synonymous* if and only if they are paired in an optimal translation of the one language into the other.

Some comments on these definitions. Translation, one intuitively supposes, is a language–language mapping that preserves meanings. On the assumption that it is analytic sentences that embody meanings, this intuition would find more penetrating expression in the thesis that translation is a mapping that preserves analyticity. This thesis, I think, is not all that far from being correct. We need only replace the discredited notion of analyticity with the more homely notion of semantic importance, and we are pretty much home. Translation is a mapping that preserves *semantic importance*. The aim of translation is to find, to the extent that it is there to be found, an *intensional structure* in the target language that parallels (is isomorphic with) the intensional structure of one's own. Faithful translation between languages is possible only to the extent that the two languages involved do indeed have closely similar intensional structures. If they are radically different in this respect, then any "translation" of the alien tongue will amount to no more than a *calibration* of the aliens (or rather, of their verbal behaviour) conceived as instruments of measurement and detection; it cannot be construed as a guide to their own conception of things.

9. *Translation: some competing views*

The conception of translation just outlined contrasts sharply with the conception embraced by Quine in chapter 2 of *Word and Object* (M.I.T. Press, 1960). The "intensional structure" of any language is held by Quine to be radically underdetermined by the linguistic dispositions of its speakers, that is, by any evidence conceivably available to a translator, and hence its "preservation" cannot serve as a criterion for uniquely correct translation. Quine begins the discussion with what appears to be a more liberal conception of translation. As Quine sees it, translation is a mapping that preserves *dispositions to speech behaviour*; it preserves the dispositions of the speakers involved to produce, or to assent to, the sentences paired by the translation. Most importantly, it preserves dispositions to produce or to assent to the *observation sentences* paired by the translation, for it is with observation sentences, apparently, that translation must begin. For Quine, this is no mere methodological point. As it emerges more clearly in 'Epistemology Naturalized',[5] Quine wishes us to take seriously a form of "verification" theory of meaning, where the unit of meaning is the entire network of a speaker's beliefs, and where the "net empirical content" of such a unit is represented by the resulting dispositions to assent to (and dissent from) observation sentences. The proper aim of translation, accordingly, is to produce a mapping that preserves this "net empirical content". It is to produce a mapping that pairs the observation sentences of the one language with the observation sentences of the other in such a way that the dispositions of the respective speakers to assent to (and dissent from) the paired sentences coincide. (That there is always more than one overall mapping of terms that will turn the desired trick at the level of observation sentences is the thesis of the indeterminacy of translation.)

This conception of translation is an interesting one, appealing

[5] In W. V. Quine, *Ontological Relativity and Other Essays* (New York, 1969).

in the purity of its behaviourist overtones, and the natural alternative, perhaps, if one has lost confidence in the idea of a mapping
that preserves intensional structure. Insofar as this latter idea rests
on the notion of analyticity (and Quine thinks it does; see *Word
and Object*, §42), Quine's lack of confidence in it is entirely justified.
But as we have seen, the idea of an identifiable intensional structure
survives the purging of the discredited notion of analyticity, and
it is quite capable of funding a conception of translation on its own.
Moreover, the notion of semantic importance upon which this
conception rests need not violate behaviourist sensitivities. As
characterized earlier, it is innocent of essentially mentalist excesses.

None of these remarks are intended as criticism of the familiar
arguments that follow upon Quine's conception of what translation
involves. If we are limited to pairing observation sentences by
similarity of stimulus meanings (see *Word and Object*, §8), thence
to confront the broader task of pairing all the non-observation
sentences in a fashion consistent with this partial mapping, that a
walloping indeterminacy should be the result seems to me not at
all surprising. What I do wish to deny is that the available empirical
constraints on translation, properly conceived, are as minimal, *or
as maximal*, as Quine's conception of translation would make
them. More bluntly, I wish to reject Quine's conception of what
translation is.

First and foremost, the primacy Quine assigns to the pairing of
the observation sentences in each language seems to me to be
profoundly misplaced. For it is not difficult to imagine a pair of
linguistic communities across which translation is entirely unproblematic, where translation is homophonic even, but where the set
of sentences *observational* for the alien community is almost
entirely without translational analogues in the set of sentences
observational in the home language. We need only imagine a community that has succeeded in making the systematic shift in their
conceptual and perceptual habits discussed at length in §4. There,
it will be recalled, a community purged itself of most of our ordinary
observation vocabulary, and learned to speak, think, and *observe*
in the terms of modern physical theory. Consider now the matter
of translating their language as if it were a foreign language. If we
are guided here by similarities in stimulus meanings (roughly, the
range of stimulus conditions that prompt assent to a given sentence),
we shall end up pairing 'This is disposed to reflect EM waves

selectively at 0.63 µm' in their language with 'This is red' in ours; 'This has a high mean molecular KE' in their language with 'This is hot' in ours; and so on and so forth. Without pausing to explore the horrors this translation would involve when we try to pair up the terms and sentences of the two languages generally, it is plain that such a "translation" would perpetrate a violent misconstrual of what is going on in the more advanced community. By contrast, there is an unbeatable translation of these foreign observation sentences into their *homophonic* analogues in the home language, analogues which, as it happens, are currently *non*-observational for us. Here we have an ideal translation in which the pairing of sentences observational in each language is negligible and incidental. The observation sentences of the foreigners find their proper translation in the highly theoretical sentences of the home language. Why Quine's conception of translation should compel our assent in the face of cases such as this is far from clear.[6]

To make matters worse, we can contrive cases in which translation is homophonic and unproblematic in any ordinary sense, but where the speakers across the linguistic divide have radically different sense organs, and hence share not even so much as a stimulus meaning in common with us. Consider a race speaking what translates smoothly as "scientific English", a race whose observational subset of the relevant expressions mirrors the very different aspects of reality to which their alien sensory equipment (radar ears, electron microscope eyes, magnetometer hands, gas-chromatographic skin, etc.) gives them access. Here again, it is the highly theoretical sentences of the home language that demand pairing with the foreign observation sentences, but here there simply *are no* domestic observation sentences sharing stimulus meanings with anything the foreigners produce. Here it is not merely unadvisable to translate as Quine's picture demands; it is quite impossible. And yet the "foreign" language may translate smoothly even so.

What this means is that the aims of translation should include no fundamental interest whatever in preserving observationality. Whether a foreign sentence is observational or no, and whatever its stimulus meaning happens to be, has nothing essential to do with how that sentence should be translated into the home language.

---

6 The relevance of such cases to Quine's conception of translation was pointed out to me by Patricia Churchland.

More generally, it means that languages, and the networks of beliefs they embody, have an identity that transcends and can remain constant over variations in the particular sensory conduits to which they happen to be tied, and in the particular locations within the language where the sensory connections happen to be made. Accordingly, any conception of translation that ties its adequacy to the preservation of "net empirical content" as conceived by Quine will lead to nothing but confusion.

An alternative view of translation, still in the behaviourist spirit, is the view that the fundamental aim of translation is just to preserve the dispositions, on the part of the speakers being translated, to assent to (and dissent from) the sentences paired in the translation. Here no special significance is assigned to observation sentences; it is sentences in general that command our attention. As Donald Davidson remarks, reflecting on the nature of translation,

> We make maximum sense of the words and thoughts of others when we interpret in a way that optimizes agreement... [7]

This is commonly known as 'the Principle of Charity' since it bids us translate foreign sentences, so far as is possible, in such a fashion that those accepted by the foreigners are paired with domestic sentences we count as *true*.

This view, I think, is also mistaken, but not by much. In fact, if we defocus my concept of semantic importance so that it reduces to its principal component – universal acceptance in the relevant group – then the view of translation proposed earlier collapses into Davidson's view. Since I have proposed that

> An optimal translation is a mapping that preserves *semantic importance*,

the assumption that

> semantic importance = universal acceptance

would commit me to the conclusion that

[7] Donald Davidson, 'On the Very Idea of a Conceptual Scheme', in *Proceedings and Addresses of the American Philosophical Association*, vol. 67 (1973–74). See also his 'Belief and the Basis of Meaning', *Synthese*, vol. 27 (1974), and 'Thought and Talk', in *Mind and Language*, ed. Guttenplan (Oxford, 1975).

An optimal translation is a mapping that preserves *universal acceptance.*

And this is clearly very close to Davidson's view that the best translation is a mapping that *maximizes agreement.*

Thus the similarity. But it is the differences here that count. Semantic importance, as that which is preserved in faithful translation, consists in more than mere universal acceptance. As explained earlier, of the sentences meeting this initial criterion, greater semantic importance will attend the general over the singular sentences, the more frequently used or presupposed over the inferential wallflowers, and those most commonly used as term-introductory formulae. In the case of a translation that actually succeeds in pairing the set of universally accepted domestic sentences with exactly the set of universally accepted foreign sentences, the relevance of these finer-grained factors will not be particularly apparent. Their relevance will emerge more clearly, however, if we consider cases where there does not exist a translation that pairs up universally accepted sentences in the manner described. In such cases we shall be forced to choose (if we wish to translate at all) from a variety of possible term-mappings, all of which pair *some* of the sentences universally accepted in each community with sentences that are problematic or widely denied in the other.

In such cases, Davidson's view counsels that we translate so as to minimize this unavoidable disagreement, that we embrace the term-mapping that maximizes congruence of universal acceptance. But why accept such a view? The defence supplied by Davidson consists in pointing out that any successful translation must rest on a presumed core of shared convictions, since genuine *dis*agreements can be identified in the first place only against the background such agreement provides. Let us agree that this is so, that if translation is possible at all, it must find some non-trivial core of agreement across the linguistic divide. Clearly, however, this modest conclusion does not by itself entail the much stronger claim that the *best* translation is always the one that finds the *most* agreement across the divide.

There is, moreover, positive reason to doubt this stronger claim, for there are factors other than raw agreement to be considered here. For example, the internal *consistency* of the set of beliefs that a given translation requires us to assign to the aliens is surely at least as important a measure of translational success as is the extent to which the alien opinions get represented as coincidental with our own. And the same point holds for the inductive and explanatory *coherence* that a proposed translation finds in their assembled convictions. If we are bent on being charitable, surely it is more reasonable to concede the aliens a normal endowment of deductive and inductive talents, and to assume that whatever their beliefs may be, they at least form a more or less coherent story, false or unorthodox though it may be. The aim of translation, after all, is to maximize the extent to which we understand *them*, not to maximize the extent to which they *agree* with *us*. In short, some agreements are more significant than others, and simply maximizing agreement at the expense of ignoring these differences can result in serious misrepresentation of the aliens' language and convictions.

If we are going to effect a cross-language mapping of terms then, I suggest we must take special care to preserve assent/dissent dispositions with respect to the *semantically important* sentences containing those terms, and the preservation or non-preservation of speech dispositions beyond these can be left to come out as they may. Of two competing mappings (i.e. sets of analytical hypotheses), one of which in fact preserves semantic importance, and the other of which achieves a somewhat greater overall preservation of speech dispositions at the expense of failing to preserve semantic importance, the latter is a translation that will fail of success in other dimensions. The internal consistency and coherence displayed by the aliens' convictions as latterly translated will be less than that which emerges from the former translation, for the set of sentences semantically important in the aliens' language forms the (presumably coherent) essentials of their collective view of the world, a set with which the rest of their beliefs will presumably also be consistent and coherent, a set of beliefs on whose appreciation by us rests the possibility of our appreciating the *rationale* of their overall system of beliefs. Correlatively, the fluency and effectiveness of *argument* across the linguistic divide will suffer seriously at the hands of a translation that fails to preserve an intensional structure that closely parallels our own.

These considerations, it must be admitted, are not so strong as they would be if they rested on a real analytic/synthetic distinction – they are only as strong as the notion of semantic importance – but the point is non-trivial even so. The aliens' beliefs and dispositions to make observation statements may differ substantially from ours, even if the intensional core of their language closely parallels our own, and there can be no point in exaggerating our agreement on matters generally if it fouls the translation in the other respects cited. Davidson, to be sure, wishes to allow for the possibility of disagreement, even extensive disagreement, and it would be misrepresentation to pretend otherwise. But the maximization of agreement remains the cornerstone of his account, and this, as we have seen, must be qualified. We do not necessarily make maximum sense of the words and thoughts of others when we interpret in a way that simply optimizes raw agreement.

With these considerations we have returned to the conception of meaning and translation proposed in §8. The meaning of a term in L is to be construed as the set of sentences semantically important for that term in the community of L-speakers. And the aim of translation is a term–term mapping that maximizes the preservation of semantic importance. That conception, it will be recalled, was introduced as the general form of a solution to the problem of term synonymy across languages. There was a similar problem, however, with the notion of term *homodoxy*, or sameness of individual understanding across idiolects. The solution to that problem, I think, takes the same form, and for much the same reasons. Consider the following definitions (here confined, for the sake of simplicity, to the case of idiolects sharing a common syntax).

(iii) An *optimal transdoxation* of idiolect I_x into idiolect I_y is a mapping of the terms of I_x into the terms of I_y such that for each term T_x the images, under that mapping, of the systemically more important of the I_x-sentences containing T_x are exactly the systemically more important of the I_y-sentences containing its paired term T_y.

That is, transdoxation should preserve *systemic* importance. With this idea in hand, a suitable definition of homodoxy is easily framed.

(iv) Two terms, T_x in I_x and T_y in I_y, are *homodoxical* if and only if they are paired in an optimal transdoxation of the one idiolect into the other.

In the case of idiolects it is maximally clear that an indiscriminate attempt to preserve speech dispositions generally is not the proper approach. Any penetrating appreciation of another person's understanding must involve an appreciation of which of the sentences he accepts are explanatorily and epistemologically most crucial in his particular view of reality. Our ability to anticipate and comprehend much of his *intellectual* behaviour – the arguments he gives and finds convincing from others, the explanations he finds plausible, the manner in which his beliefs change under the pressure of new information – will depend heavily on our having an appreciation of this more discriminating kind. On questions of homodoxy then, especial weight must be assigned to those sentences systemically important for the individuals involved. That is, idiolects have a systemic structure – analogous to the intensional structure of a language – and accurate transdoxation is possible only to the extent that the idiolects involved share closely similar systemic structures. In the absence of such similarity, the people involved will find penetrating discussion on any but the most mundane of topics a considerable struggle. The phenomenon, I presume, is familiar. In the following section, I shall address the problems it raises.

Finally, we must consider a difficulty from an unexpected quarter. In 'The Meaning of "Meaning"' (pp. 131–193) Hilary Putnam has argued that any account along the lines here given must be fundamentally inadequate. I shall now try to outline the relevant arguments and meet the criticisms they imply.

Putnam is concerned to drive a substantial wedge between the meaning of a given term, and the set of beliefs involving that term embraced by any arbitrary speaker. So far, of course, we agree. My account distinguishes sharply between a given individual's understanding of a term, and the meaning of that term in his language. For me, meaning is an objective, collective, and occasionally even a quasi-official matter. But while Putnam might be sympathetic towards my account of meaning construed as an account of *intension*, he would certainly reject it as an account of *meaning*, and he would deny that the preservation of such intensions is the primary aim of good translation.

His reason is straightforward. The extension of a term – of a natural-kind term, at least – is not fixed or determined (primarily?) by its intension. Rather, its extension is fixed in a quasi-demonstrative fashion by way of an indexical formula like 'Something is in the extension of the term "water" if and only if it bears the relation "same liquid" to the stuff we call "water" around *here'.* A direct consequence of this position is that a pair of terms, each from a distinct linguistic community, could be identical in their intensions and yet completely different in their extensions.

The principal example that Putnam constructs in illustration of this position concerns the existence and discovery of a planet we may call Twin Earth, a planet indistinguishable from Earth save for the microstructure of that liquid which there doubles very faithfully for water. More to the point, the aliens' *beliefs* concerning this substance (whose chemical formula is not H_2O, but rather something XYZ) match our own beliefs about water in every detail. That is, the respective intensions of 'watter' (I shall misspell our term to indicate the Twin Earth usage) and 'water' are identical. But we cannot translate 'watter' by 'water', argues Putnam, because 'watter' has as its extension that scattered liquid found on Twin Earth, and that liquid is simply not water, since it does not stand in the relation 'same liquid' to the stuff we call 'water' around here. It is not H_2O, but XYZ. Equally, and for the the same reason, our stuff is not watter. Accordingly, 'watter' and 'water' do not have the same meaning. They have the same intension to be sure, but different extensions. We must therefore construe extension as an independent component of the vector we call meaning, a component coequal with intension.

A second case of importance for Putnam concerns a putatively common extension for two terms with conflicting intensions. Putnam argues that the extension of 'χρυσός' in Archimedes' dialect of Greek is the same as the extension of 'gold' in modern English, despite incompatibilities between the respective intensions of those two terms. And the reason is that the relevant instances of the indexical formula (see above), as used by Archimedes and as used by us, pick out the very same natural class – things composed of Au atoms.

Putnam's general position here gives expression to what he describes as a strongly realistic attitude towards those objective and natural classes for which our current terms, or rather their intensions, may provide only a partial and approximate conception. By way of indexical formulae like the one above, our speech is alleged to be firmly "attached" to those natural classes despite confusions and differences in our intensions, differences that may reflect theoretical conflict at a given time, or theoretical change over time. In sum, while our vision of those natural classes may be confused and out of focus, we can all have and maintain a rigid *grasp* on them while our vision slowly converges to clarity.

The intuitions to which Putnam's account appeals are largely genuine, but I think that the account itself is substantially overdrawn. We may agree with the idea that there are important, objective, natural classes "out there", classes awaiting adequate comprehension within a framework of general beliefs. But we need not agree that these classes are or can be uniquely secured as extensions beforehand, nor that such extensions as do get secured are secured by extra-intensional means. The intuitions that render Putnam's account plausible can be explained on assumptions less dramatic than these.

Let me begin by agreeing with Putnam that the meanings of natural-kind terms can and often do have a non-trivial indexical element. To concede this, however, need not be to concede that meaning has an extra-intensional component. If we construe an intension (meaning) as a set of semantically important sentences, as the account in §8 invites us, then we can make contact with Putnam by pointing out that semantic importance is not confined to general sentences. Singular sentences can and often do have substantial semantic importance. To take an example close at hand, the sentence 'The stuff in our lakes, rivers, and oceans is water' clearly enjoys a great deal of semantic importance, in all of the dimensions cited on pp. 52–3 save isolation and simple generality. Accordingly, it is part of the intension of our term 'water', and if it makes some contribution towards delineating the extension of 'water', this is perfectly normal and should not be construed as an extra-intensional means of specifying extension.

It would be unfair to object that I am here attempting to undermine Putnam's account by stipulating *ad hoc* that the relevant indexical sentences are after all part of a term's intension. For the point is neither stipulatory nor *ad hoc*. It is a straightforward consequence of the general theory outlined in §§7 and 8, a theory embraced on wholly independent grounds.

Of course, a singular sentence like 'The stuff in our lakes, rivers, and oceans is water' provides at most a sufficient condition for being in the extension of 'water', and herein lies the rationale for the second aspect – the inductive clause, as it were – of Putnam's indexical formula. Something is water if (and only if) it has the *same nature* as the stuff in our lakes, rivers, and oceans. That is, the extension of whatever term is at issue is claimed to be closed under the relation 'has the same nature as'. By way of that relation then, if we can specify any example of the relevant extension, we can specify the whole extension.

As with the first condition, I think, such significance as is enjoyed by this second condition can be accounted for by whatever semantic importance is enjoyed by the relevant instance of 'Something is φ if and only if it has the same nature as the φs around here'. But here I would object that our collective convictions in this regard are nowhere near so sophisticated and visionary as they become in Putnam's hands. 'Same nature', after all, *need* mean nothing more than 'shares the same universal features that make up the current intension of "φ"'. And this will not supply the bite that Putnam seeks. But if we are to read a deeper significance into the phrase 'same nature', surely that significance will vary from period to period in any case. Depending on what stories our scientific or metaphysical theories then tell us about the fundamental ways in which things can have the same nature, a condition of the sort at issue will pick out one extension in one intellectual period, and a quite different extension in another. For an alchemist, 'has the same nature as the gold around here' came to something like 'is ensouled by the same uniquely noble spirit as the gold around here'. And this relation not only fails to pick out the extension of our 'gold', it fails to pick out any extension at all.

Let me summarize. According to Putnam, sentences of the form 'The stuff around here is φ', and 'Something is φ if and only if it has the same nature as the φ around here', constitute an extra-intensional device for securing a uniquely natural extension for 'φ', an extension that remains constant despite possible changes/differences/evolution in the intension of 'φ'. According to me, such sentences are just run-of-the-mill elements *within* the intension of 'φ'; their contribution to the determination of its extension is real, perhaps, but unexceptional; and they are as likely as any other elements in that intension to suffer qualification, reinterpretation, or outright rejection in the face of new discoveries and improved understanding. On this construal, they may occasionally provide a relatively stationary point around which other elements may shift, but they do not constitute a permanent pier that will forever anchor the boat of extension against the shifting tides of intension: they are themselves a part of the fickle tide.

If we embrace the alternative position just outlined, it is plain that we may still agree with much of what Putnam has to say about his principal example. Specifically, we can agree that 'watter' and 'water' do not have the same meaning and should not be paired in translation. But our reason will be that these two terms display what happens to be an important difference in intension. To begin with, the reference of 'The stuff in our lakes, rivers, and oceans' as it occurs in their mouths is numerically distinct from its reference as it occurs in ours. Of itself, this difference need not preclude translating 'watter' by 'water', since the two sufficient conditions (being in their lakes, and being in ours) may be wholly compatible, and their non-identity need reflect nothing more substantial than the non-identity of our respective locations in space. But in this case the difference happens to be very important: given that the stuff in the aliens' lakes is not H_2O, but rather XYZ, *the intension of their term 'watter' is empirically incoherent.* For *ex hypothesi* that intension contains both (1) 'The stuff in our lakes, etc., is watter', and (2) 'Watter is H_2O', which elements conjointly entail an empirical falsehood. The aliens are thus faced with the choice (or will be, when enlightened about XYZ) of counting either (1) or (2) as false. And so are we, since the manner in which we translate 'watter' must anticipate that choice. To translate 'watter' by 'water', for example, would be to construe them as being right about (2), but mistaken about (1). But almost certainly *they* would insist upon the truth of (1), and concede error on (2), since the disruption of their day-to-day speech habits will be very much less on this alternative than on the other. Accordingly, we must not translate 'watter' by 'water', since the only coherent intension we can reasonably anticipate for 'watter' is (would be) significantly different from the intension of 'water'.

In this case then I do concur with Putnam's conclusion, though for reasons that involve no appeal to his second condition. In most of the other cases he considers, however, I would be inclined to dispute his conclusions, since those conclusions depend heavily on presumptuous applications of his second condition.

These brief remarks do not do justice to the many careful examples that Putnam constructs. But a complete discussion of them is impossible in the present context, and I will rest content with having outlined a different way of approaching them.

10. *Communication and commensurability*

Commensurability, or rather *in*commensurability, has become the focal point of much discussion recently. The structure of the problem and the nature of the worries are readily outlined.

'Suppose all of the sentences a man accepts to be of non-zero semantic relevance, to contribute something to the meaning of whatever terms they contain. Then the slightest change in the set of sentences he accepts would amount to a change (for him) in the meanings of the terms immediately involved, and hence in the meanings of all other accepted sentences containing those terms, and hence in the meanings of the further terms they in turn contain, and so on. Correlatively, even the smallest differences in the sets of sentences embraced by two different speakers would entail systematic differences in the meanings of all the terms they share. It would then be a mystery how they ever understood one another, or what sense we could assign to the notion of disagreement between them. For any sentence S to which they assent and dissent respectively, the difference in meaning S has for each precludes our characterizing their "disagreement" as a case of mutual contradiction. That is, their respective conceptions or accounts of the world would have to be classed as incommensurable since, on the assumptions made above, there would appear to be no logical contact between them. They would be different accounts, presumably, but not logically incompatible accounts. It would not even be clear in what sense they are competing accounts.'

This, in a nutshell, is the sort of "semantic solipsism" to which many fear a holistic account of meaning must commit us. Nor of course do the worries end here. Given the logical isolation that the proponents of different theories are supposed to suffer, the possibility of comparative evaluation across theories becomes highly problematic, as does the notion of progress through a series of theories over time. Incomparability seems everywhere, and a thoroughgoing epistemological relativism seems the unhappy but inescapable result.

Surely indeed, the true situation is not so desperate. However, the route to reconstituting our faith in relevant communication and effective comparison lies not in rejecting the holistic approach to meaning and understanding. It lies in learning to take it seriously. The fact is, incommensurability is a matter of degree, and we live with it every day. In subtle degree it is a background feature of almost all of our non-trivial discourse with one another. And yet it troubles us little or none at all. The proper approach here, I suggest, is not to try to paper over these difficulties with some less sensitive theory of meaning contrived especially to prevent their ever arising. The proper approach is to concede the reality of these difficulties, and then ask why these difficulties do not in fact trouble us to anything like the degree that the preceding *reductio* suggests they should. In hopes of making good on these remarks, let me return to the conceptions of meaning and understanding outlined in the preceding sections, the immediate aim being to provide a more finely resolved image of what the difficulties involved here come to.

The first point to be made is that the account of meaning and understanding outlined in §8 is not "holistic" in the truly radical sense on which the *reductio* rests. On the account proposed in §8, meaning and understanding are determined by the semantically and systemically more important sentences, respectively, rather than by the set of *all* accepted sentences. Divergence of meaning (or understanding) need be conceded then, only to the extent that there are differences in the sets of sentences semantically (systemically) important for the term at issue within the relevant groups (individuals). The spectre of an inevitable semantic solipsism, therefore, may safely be put aside.

What we do have to come to grips with here is the somewhat less dramatic phenomenon of semantic (and systemic) divergence, a divergence that comes in degrees, ranging from the negligible to the unbridgeable. Both semantic and systemic importance, it will be recalled, were matters of smooth degree, and in consequence both the sameness of meaning across groups and the sameness of understanding across individuals must be counted matters of degree as well. If we are bent on tracing the meaning of a term into a certain cluster of sentences containing it, and the meanings of the terms they in turn contain into further clusters of sentences, and so on, then it is clearly the entire network such tracing discovers that forms the basic unit of pairing in cases of translation. And the problem is that perfect isomorphism or exact identity of intensional structure between natural languages (or of systemic structure between idiolects) is not quite what the real world shows us. Close similarities, yes. Perfect identities, no. That is, few if any pairs of natural languages admit of a mutual translation that is optimal in the sense defined earlier. And perfect identity of systemic structures across idiolects is, if anything, even less likely than its social parallel across languages, for systemic structures tend to be rather larger and more intricate than the comparatively gross intensional structure of whatever language they instance. There is therefore greater scope for subtle divergence among idiolects.

What all this means is that the synonymies we standardly see across languages, and the homodoxies we standardly see across idiolects, are very seldom the perfect synonymies or homodoxies we might suppose them to be. But I see not why this should trouble us overmuch, either as speakers who must deal with the problems such divergence may occasionally introduce, or as philosophers who must theorize about those problems. The fact is, such failures of perfect correspondence are, for the most part and for most concerns, of no practical significance, and linguistic commerce will generally proceed most efficiently on that very assumption. And when on some topic divergences do rise to the surface and do prove significant for the point under discussion, the speakers involved can usually fall back, as speakers do, on the extensive substructure of meaning and understanding that remains shared between them, and then approach the troublesome topic anew from that common foundation, arguing from that foundation as they go, and becoming "bilectical" in the process as they come to learn and appreciate the divergent intellectual structure embraced by their opponent. If the discussion is successful, the differences between them get hammered out, either by one speaker being won over to the systemic (or semantic) structure of the other, or by their common conversion to a compromise structure forged in the course of the debate. The discussion, of course, may also be unsuccessful, and the speakers may thereafter avoid one another as having weird or impenetrable views.

All of this, I am sure, is familiar. Though we are not the victims of a perennial Babel, penetrating discussion very often is a struggle, for the reasons and along the lines indicated. Why object to the account of meaning and understanding proposed then, on grounds that it commits us to the existence of such difficulties? Since such difficulties are both real and common, if usually minor, the account would be inadequate if it did not acknowledge them.

On the other hand, the account proposed does raise some serious

theoretical questions, if not in the form of objections to the account
itself, then as major puzzles that none the less want solving. What of
communication and evaluation across alternative conceptual frameworks where the differences are not minor and localizable, but major
and systematic? What of cases where there is no isolable substructure
that is both genuinely shared by the thinkers involved and sufficiently extensive to serve as a useful platform from which their
differences can be approached and arbitrated? In short, how can
intellectual commerce proceed in cases of *radical* incommensurability?

I would like to suggest that the process is identical, at bottom,
with intellectual commerce in normal cases. The first requirement,
of course, is that one *learn* the intensional structure of the alien
framework, or the systemic structure of the divergent idiolect. (In
normal cases, by definition, this requirement is already met: the
framework of the other speaker, the bulk of it at least, is also one's
own.) Argument and criticism can then be conducted within the
other framework in unproblematic fashion. Also, the other speaker
can learn one's own framework, therein to perform the analogous
critical function. The result is two simultaneous internal evaluations
of two comprehensive alternatives competing for our commitment.
And the choice is made not on the relative happiness of the epistemic
relations they bear to some specific convictions we both *share*
(a set of observation sentences, for example), for in the radical
case at issue we share no such thing. The choice is made rather on
grounds of the relative "internal" virtues of the two alternative
frameworks: on their inductive coherence, their explanatory unity,
their informational richness, and suchlike (somehow understood).

The important thing to note about this approach to cross-
theoretical comparison in cases of systematic divergence of principle is that it places such comparison on a straight continuum with
cross-theoretical comparison in cases of minimal (peripheral)
divergence. There, too, the evaluation is made basically between
two comprehensive alternatives: the large set of commitments we
can be presumed to share plus the first of the two theories at issue,
versus the same set of shared commitments plus the second of the
two theories at issue. Schematically, B & T_1 versus B & T_2, where
B is "shared background commitments".[8] In such cases it may

---

[8] Generally speaking, the elements of B are "common" to both alternatives only if
conceived in abstraction from their interconnections with T_1 in the first alternative

appear that we are evaluating T_1 and T_2 against the stable standard that B provides. And so we are, in the sense that the alternative that embodies the best B/T fit will be our choice. But the alternative that embodies the best B/T fit is just the alternative that maximizes the *internal* virtues of overall inductive coherence, explanatory unity, and so on. That is, the choice between them is made on the same kinds of grounds, ultimately, that govern our choice in cases of radical divergence. The only difference between the two ends of the spectrum is the amount of material the pairs of global alternatives *have in common*. A great deal of shared material may make both the evaluation and the choice much easier to make, but it does not make either the evaluation or the choice any different in kind. Choices between widely divergent or "incommensurable" alternatives, therefore, are not so different after all. They are merely somewhat unnerving in the sheer magnitude of the choice they present.

The main burden of the preceding discussion has been to undermine the dichotomy popularly drawn between (a) epistemic alternatives that are commensurable, and (b) epistemic alternatives that are *in*commensurable. Cases of each kind now appear as no more than the extremal points of a smooth continuum of cases. The vast majority of actual epistemic alternatives will fall very near the first extreme, to be sure, but that is only because the vast majority of epistemic alternatives are highly prosaic and without notable consequences for one's intensional or doxastic system, whichever alternative happens to be chosen. By the same token, they are the least interesting of the epistemic choices we confront. The more interesting choices, by contrast, are precisely those where part of what is at stake is the possible need to rethink (revise) something of current systemic and/or semantic importance, and the more interesting they are, the farther up the spectrum in the direction of "incommensurability" the relevant alternatives will be.

Now what this means, from the point of view of epistemological theory, is that an adequate general account of the way in which rational evaluations and choices are made cannot be predicated on

---

and with T_2 in the second. If those elements are conceived (respectively) within those larger contexts, their semantic and/or doxastic identities may differ substantially. In the sort of case at issue just here, however (the differences between T_1 and T_2 being without systemic significance), the schizophrenia suffered by B will be negligible.

assumptions that are appropriate only at the limit of one end of the spectrum of possible cases (the uninteresting end). As honest epistemologists we must appreciate that sooner or later we are going to have to face the music. We may perhaps gain time (at the expense of losing ground) by steadfastly denying that this, that, or these familiar historical conflicts involved a clash between incommensurable theoretical alternatives, but the fact remains that situations of choice between radically incommensurable alternative theories are certainly *possible* (infinitely many, in fact). Accordingly, either we must deny outright that rational choice is possible at all in such cases, or we must accept the responsibility of constructing our normative epistemological theory so as to handle them. The first alternative is a needless concession to scepticism; it renders arbitrary some of the proudest epistemic decisions our intellectual history contains; and it clashes fatally with the thoroughgoing continuity we have discovered between the "commensurable" and the "incommensurable". Let us agree then, that an adequate account of rational evaluation and change of belief must acknowledge the entire range of possible cases, and must give a unitary account of all of them. More specifically, what it *cannot* do is to provide a general account of rational belief change that is based on a blanket assumption of stability in the background conceptual framework or language. And what it must do is account for rational conceptual change as just an extreme but nevertheless continuous case of rational belief change.

11. *Intertheoretic reduction and conceptual progress*

It happens fairly frequently, in the evolution of human understanding, that an established way of conceiving of a certain domain proves to be reducible to some new way of conceiving things. Newtonian mechanics is reduced by relativistic mechanics; classical thermodynamics is reduced by the kinetic/corpuscular theory of heat (statistical thermodynamics); and so on. This is commonly thought to be a good thing for both the old framework (the reduced theory) and the new framework (the reducing theory). It is a good thing for the old framework, it is said, since its categories and principles are thereby reaffirmed or vindicated. And it is a good thing for the new framework, it is said, since it thereby inherits whatever confirmation supported the older view.

Both of these claims are false, though there is enough truth in each to have won them a substantial following. Just how they are false, and why it matters, is the concern of the present section.

As even casual scrutiny will reveal, the reduction of one theory to another bears certain substantial similarities to the translation of one language into another. In both cases we find a mapping of one vocabulary into another, a mapping that preserves certain featues thought to be important. But it is not meaning that is preserved in intertheoretic reduction. Indeed, the pairings effected therein standardly *fail* to preserve meaning. Consider 'temperature' as paired with 'mean molecular kinetic energy', 'light' as paired with 'electromagnetic waves', and so on. Synonym-pairs these are not.

Reduction, accordingly, is not translation, but it is, I think, another species of the same genus. The difference lies in the function performed by the mapping at issue, a function that places much weaker demands on a reduction mapping than must be met in a translatory mapping. To see this, consider what an ideal or maximally smooth reduction provides us. First, it provides us with a set of rules – "correspondence rules" or "bridge laws", as standard vernacular has it – which effect a mapping of the terms of the old theory (T_o) onto a subset of the expressions of the new or reducing theory (T_n). These rules guide the application of those selected expressions of T_n in the following way: we are free to make singular applications of those expressions in all those cases where we normally make singular applications of their correspondence-rule doppelgangers in T_o. In this way, a reduction *locates* the newer theory within the conceptual space currently occupied by the older theory. It provides the basic instructions, as it were, for the orderly displacement of the latter by the former.

Second, and equally important, a successful reduction ideally has the outcome that, under the term mapping effected by the correspondence rules, the central principles of T_o (those of semantic and systemic importance) are mapped onto general sentences of T_n that are *theorems* of T_n. Call the set of such sentences S_n. This set is the image of T_o within T_n. It is not required of these selected sentences of T_n that they be semantically or systemically important as well – only that they be consequences of the central principles of T_n.

Such an outcome means two things. It means (a) that the displacement mentioned above can be an *orderly* displacement. It means that the displacement of T_o by T_n would not require substantial revisions in the larger network of background beliefs of which T_o is presumably an integrated part. Since T_n contains, as a proper substructure, a set of principles S_n that is isomorphic with the set of principles comprising T_o, we are thereby assured that T_n will *cohere* with the background in the same ways that T_o did, and will perform all the same predictive and explanatory functions that T_o performed.

Further, such an outcome means (b) that the reducing theory T_n can lay claim to confirmatory considerations systematically analogous to those that have already accrued to T_o. For consider any singular sentence in the vocabulary of T_o that is taken to be evidence in T_o's favour. Given the initial mapping effected by the correspondence rules, there must also be, for each such sentence, an affirmable doppelganger in the vocabulary of T_n, a singular sentence that bears the same relation to S_n that its twin bears to T_o. Accordingly, if T_o is confirmed, then so also is S_n, and thus (derivatively) T_n itself.

In sum, what a successful reduction shows us is that one way of conceiving of things can be safely, smoothly, and – if the excess empirical content of T_n over S_n is corroborated – profitably *displaced* by another way of conceiving of things. And this, I submit, is the function of reduction. *A successful reduction is a fell-swoop proof of displaceability*; and it succeeds by showing that the new theory contains as a substructure an equipotent image of the old.

Displacement, of course, need not actually take place, however much considerations of unity and simplicity might demand it. Familiarity, entrenchment, convenience, and continuity may together counsel a less puritan course. And in the case of an ideal reduction of the sort characterized above there should be no impediment to the claim that the items/properties paired in the correspondence rules are contingently identical with one another, in the fashion of 'light = electromagnetic waves in the 0.5 μm range', 'temperature = mean molecular KE', and so forth. The advantages of such a claim are straightforward. We thereby pare our ontology in the manner simplicity requires, and we salvage the legitimacy of a familiar idiom at the same time. For these reasons, the identification option will probably be taken wherever it can be.

Indeed, why be grudging here? A successful reduction of the ideal sort described provides an excellent reason for asserting the relevant cross-theoretical identities, the best reason one can have.

Of course, this is all a "best case" analysis: it is the ideal reduction we have been discussing. But reductions are seldom if ever ideal, and it is time to see how the ideal gets frayed and bent in real cases. Before proceeding to this, however, let me emphasize two final points about the account as given so far, since it is important to appreciate that they hold even for an ideal reduction.

First, on the account given above it is not the reduced theory, T_o, that is deduced from the principles of T_n, as some other accounts have it.[9] What is deduced from T_n is rather the set S_n, an equipotent image of T_o within the idiom of T_n.

Second, it is important to appreciate that cross-theoretical identity claims, even if they are justly made, are not a part of the reduction proper, and they are not essential to the function it performs. The correspondence-rule pairings need not be construed as identity claims, nor even as material equivalences, in order to show that T_n contains an equipotent image of T_o. In fact, we can treat each correspondence rule as a mere ordered pair of expressions – for example, ⟨'electric current', 'net motion of charged particles'⟩ – and we will then need only the minimal assumption that the second element of each pair truly applies where and whenever the first element of each is normally *thought* to apply. Such an assumption, note, is strictly consistent with the idea that the first elements (the expressions of T_o) do not apply to reality at all.

With these points in mind let us now consider reductions that fall short of the ideal described. Deviations from the ideal basically come in one or both of two dimensions. First, the image, such as it is, of T_o within T_n may not be a complete or wholly faithful image of T_o. It may be, for example, that one or two of the important principles of T_o are mapped onto sentences in T_n that are just false, while the remainder of T_o reduces smoothly. Or it may be that several of the principles of T_o must be modified or "corrected" in some way before they find an appropriate image in T_n. Here we may say that it is not strictly speaking T_o that finds an appropriate image within T_n, but rather some theory T_o' closely similar to T_o.

Second, it may be that the image, S_n, of T_o within the vocabulary of T_n is not an unaided consequence of the basic principles of T_n.

⁹ See Ernest Nagel, *The Structure of Science* (New York, 1961), ch. 11.

S_n may be derivable within T_n only if we include in the premises of the deduction some limiting condition or counterfactual assumption. (For example, to winnow an image of classical mechanics out of special relativity, assume 'For any actual $v \neq c$, v/c is negligibly small'. The result, as Feyerabend and Kuhn have stressed,[10] is not classical mechanics itself, but rather an equipotent image of classical mechanics expressed in what is still the peculiar idiom of special relativity.) If the assumption in question relevantly approximates the domain in which T_o has hitherto been successfully applied, the result will still be counted a reduction. Here it is strictly speaking not T_n that provides an equipotent image of T_o, but rather an augmented theory T_n' closely similar to T_n. And finally, some cases may instance both of these complications, T_o being related to T_n by way of some T_o' *and* T_n'.

Cases of this kind may still count as reductions of T_o by T_n, despite the warts, since they may still perform the essential function of a reduction. Specifically, they may still illustrate that T_o could be more or less smoothly and painlessly displaced by a superior theory T_n some of whose resources closely parallel (in the relevant contexts) those of T_o. The displacement may require some revisions here and there, and the parallels may not be perfect, but the degree to which the reduction succeeds may be found far more interesting and important than the comparatively minor respects in which it fails.

If we broaden our concept of reduction to include the important cases just described, it is clear that we must be prepared to count reducibility as a matter of degree. Like translation, which may be faithful or lame, reduction may be smooth, or bumpy, or anywhere in between. But it must not be thought on this account that the bumpy ones will be less valuable or less interesting than the smooth. On the contrary, they may well be as interesting for what they fail to preserve as for what they preserve, and more valuable for the revisions they dictate than for the commonplaces they leave unscathed.

Further, on the liberal account we are here embracing it is clear that a true theory may reduce a false one. For the falsity, even the radical falsity, of T_o need not preclude its having some large fragment or modified version T_o' that finds an appropriate

[10] See Paul Feyerabend, 'Explanation, Reduction, and Empiricism'; also Thomas Kuhn, *The Structure of Scientific Revolutions* (Chicago, 1962), ch. 9.

image S_n within some stoppered-down version T_n' of T_n. A true theory may contain resources that, under certain restricted conditions perhaps, parallel the resources of some false theory sufficiently closely to allow the true theory to play most of the false theory's role. And this means that a reduction is *not* essentially reaffirmative or vindicative with respect to the categories and principles of the reduced theory. On the contrary, a reduction may even pinpoint for us the way or ways in which the reduced theory is cockeyed. We need only examine the details of its deviation from an ideal reduction: the assumptions that had to be made to squeeze a suitable S_n out of T_n, and the points in which T_0 differs from the closest legitimate S_n we could find for it.

Further still, it is possible on this account for a theory to reduce even an *incommensurable* competitor. And that is fortunate, since there is no other way to describe adequately the case of classical mechanics (CM) and the special theory of relativity (STR). These two theories are mutually incommensurable in the straightforward sense that it is not possible to express (translate) the principles of either theory within the vocabulary of the other. The difficulty in this regard arises from the fact that the familiar notions of CM – mass, length, time, and the entire hierarchy of further notions derivative upon these – have no meaning in STR save as they are made relative to some reference frame. $Mass_{CM}$, for example, is an intrinsic feature of an object, but $mass_{STR}$ is a *relation* holding between an object and countless reference frames, a relation having a different value for each of those frames. And in general, where the vocabulary of CM contains a one-place (or n-place) predicate, the vocabulary of STR contains a two-place (or an $n+1$ place) predicate. The systematic differences in meaning across this pair of theories in particular is sufficiently great to show itself in a systematic difference in the syntactic configurations of the predicates involved. In vain would we try to translate back and forth.

None the less, STR reduces CM, in the sense articulated here, and in the pages of numberless undergraduate texts. For despite their incommensurability, it is easy to deduce within STR (given the limiting assumption about velocities cited earlier) a highly convincing image of CM. It is in fact so convincing that most people seem to think it *is* CM. But it is not. If the expressions in that deduced image are articulated properly – as usually they are not – they will still have one argument-place too many to be expressions of CM. By virtue of the limiting assumption, those extra places will have been rendered impotent so far as affecting the *value* of the $mass_{STR}$, or $length_{STR}$, or whatever, is concerned (that is why the image behaves like CM), but the extra places are still there and the image remains distinct from CM.

In the example of CM and STR we have a particularly striking counterinstance to the claim that a reduction reaffirms the reduced theory. For in this case it is not just the laws of CM that STR bids us throw overboard as illusory, but also the fundamental categories with which CM interprets the world. Those categories, no less than those principles, turn out to be illusions born of a parochial experience.

I do not mean to suggest, of course, that in all but an ideal reduction the categories of the old theory must suffer dissolution. A reduction need not be ideal in order that cross-theoretical identity claims be made and sustained. In the case of an imperfect reduction one might fairly assert such identities, despite certain cross-theoretical clashes in principle, by pleading error in our old beliefs about one party to the identity. The plausibility of such a move will be inversely proportional, however, to the number and importance of such old principles as the identity claim requires us to relinquish. And in the case of CM and STR there is little room even to consider it, for we would be trying to identify unary properties with binary relations.

These considerations show that we should draw a distinction between the reduction of one theory to another, and the reduction of one ontology to another. A reduction of T_o to T_n will herald a reduction of T_o's ontology to T_n's ontology only if the former reduction is sufficiently smooth to allow the relevant cross-theoretical identity claims to be sustained.

There is a final lesson displayed with unusual clarity in the CM/STR reduction, and it concerns the claim that a reducing theory T_n inherits the evidence accrued to the reduced theory T_o. In fact, T_n does not generally inherit such evidence: in particular, it does not inherit evidence for T_o expressed in any vocabulary peculiar to T_o. The supporting evidence that T_n (via S_n) does find here is rather the correspondence-rule *image* of the evidence for T_o, and the sentences in that image may differ substantially in meaning from the sentences that support T_o. Correspondence rules, it will be recalled, do not in general preserve meanings. What counts as evidence for T_o, therefore, may appear as so much gobbledygook from the point of view of T_n, even though the latter theory reduces the former. And this is precisely what we find in the case under discussion, where the classical categories in which the evidence for CM is expressed are undermined completely by the incommensurable vision of STR. The new theory may inherit the "experience" accrued during the tenure of the old, but it is an experience that must be reinterpreted before it can supply any evidence for the new.

This example, and the point it illustrates, must be taken very seriously. For it constitutes a real example of what we discussed in the preceding section: a case of two incommensurable conceptual alternatives on a scale so large that each provides a distinct interpretation of the most basic terms we use for describing reality in general and what we observe and measure about it in particular. The evaluation of such alternatives, accordingly, requires of us a kind of studied schizophrenia, and even our observational experience becomes systematically ambiguous. When we perform certain familiar operations on a speeding charged particle, have we measured its $mass_{CM}$, period? Or have we measured its $mass_{STR}$-relative-to-our-reference-frame? When we carefully determine the distance between two mirrors (as in the Michelson–Morley apparatus), have we measured the $distance_{CM}$, period? Or have we measured the $distance_{STR}$-relative-to-our-reference-frame? Or have we perhaps not measured anything very much at all, since – as on the transitional views of Fitzgerald and Lorentz – our metre sticks are absolutely shrinking and expanding in self-defeating ways?

The resolution of such ambiguities does not proceed by appeal to any simpler or more neutral set of data, observational or otherwise. It proceeds by determining which of the two conceptual alternatives allows us to construct a consistent and coherent account of our experience as it is pressed farther and farther into unfamiliar domains. In the present case the unfamiliar domain is the realm of very high relative velocities, and it is here that the classical modes of description and explanation come manifestly unstuck. The observations and measurements made faithfully in its terms according to standard procedures begin detectably to clash with one another, to refuse to cohere, and to be inexplicable within the classical compass. STR, on the other hand, functions as smoothly and coherently here as it does for low relative velocities, and is embraced on that account as the superior vehicle of understanding. The choice then, is a global choice, and it is made on what might vaguely be termed the "internal" features of the comprehensive alternatives at issue, alternatives that include many observation statements as well as the theory proper.

The guiding concerns of this chapter have been, broadly speaking, semantical, but the central lessons of the discussion in the end turn out to be epistemological. One might sum them up by saying that what an adequate epistemology must begin by acknowledging is the thoroughgoing *plasticity* of human understanding, and among the things it must be able to account for is the rationality not just of minor but of wholesale changes in the form that understanding assumes. Just how serious that challenge is will become clear in chapter 5.

4

Our self-conception and the mind/body problem

12. One's knowledge of other minds

We have before us so far a broad epistemological thesis and the outlines of a semantic theory. We must now assess the consequences of these closely related positions as they bear on the philosophy of mind. In particular, we shall see what light they throw on the nature of our self-conception generally, and on the specific knowledge one has of oneself and others, *qua* persons. We shall find, I think, that the illumination is considerable.

There is a further reason for turning, at this point in the essay, to the philosophy of mind. The epistemological theses advanced to this point fall well short of a comprehensive theory of knowledge, since they do not include a solution to the methodological problem. Beyond some casual mutterings about coherence, explanatory power, and the like, no attempt has been made to explicate what *rationality* consists in, as it bears on theoretical evolution in general. The reason is simple. I am of the opinion that this fundamental problem will require for its solution, or even for its significant advancement, a revolution in our self-conception. I must therefore establish that there is *room* for a revolution in our self-conception, that our self-conception is as speculative as any other. Once this has been established, we can explore the shape of the revolution with confidence.

It has been argued at length that our common-sense conceptual framework for empirical reality is in all relevant respects a theoretical framework. I wish here to emphasize and explore that claim as it applies to that subframework comprehending the notion of a person. Despite a long tradition of a contrary bent, I think we shall find that the theoretical nature of common-sense is more strikingly apparent here than perhaps anywhere else in our conceptual field. Let me try to make good on these claims, in stages. A fruitful place to begin the discussion is with the problematic epistemology of second and third-person ascriptions of psychological states, that is, with the traditional problem of other minds.

The problem arises simply enough. Any ascription of specific psychological states to another – of thoughts, perceptions, desires, or pains – must be made on the basis of overt bodily behaviour. And specific states are inferred from specific behaviour in accordance with certain general correlations assumed to hold between them (as between pain and grimacing, for example). But an appreciation of the general correlations that hold between such "inner states" and their "outer criteria" cannot be derived from one's observation of those correlations in others, for only the "outer" elements of the putative correlations are open to one's inspection. Knowledge of any such correlations can therefore be derived only from one's *own* case, for here, it is said, the inner elements are open to inspection. That those same correlations discovered in one's own case also characterize the cases of every other human can therefore be inferred only by way of an inductive leap from the necessarily solitary instance of one's own (possibly atypical) case. The possible justification available for beliefs about the minds of others is therefore at the vanishing point. One is limited to an incorroborable generalization from a single case. In this way do we arrive at scepticism with respect to the content and even the existence of minds other than one's own.

The attempts to avoid these awkward conclusions have traditionally taken one of two forms. The first tries to deny the need for justifying our belief in the general correlations at issue by denying that they are empirical in character to begin with. This approach is characteristic of those views variously called 'logical', 'philosophical', or 'reductive' behaviourism. The profound difficulties involved in trying to deny the empirical character of these correlations has always provided encouragement for the second approach, the argument from analogy. Here the attempt is made to show that the problematic inference from (argument from analogy with) one's own case is after all not so feeble or audacious as it might seem. Here also the battle has been uphill, but I think the main problem with this approach is not really that it has failed to represent the 'inference to other minds' as sufficiently muscular to explicate our pre-philosophical convictions. The problem with the argument from analogy is that it accedes in the representation of one's knowledge of other minds as being essentially parasitic on one's knowledge of one's own mind. Behaviourism was mistaken in denying the empirical character of "psycho/physical" generalizations, in denying that they stood in need of empirical justification at all. But it is equally wrong, I suggest, to insist that such justification can flow only from an examination of one's own case.

A moment's reflection here will reveal an available mode of justification in which the facts of one's own case are entirely superfluous. Consider again the set of generalizations whose justification is at issue – the set of sentences descriptive of the general relations holding between (a) types of causal circumstances and types of psychological states (hereafter: *P-states*), (b) the various types of P-states themselves, and (c) types of P-states and types of overt behaviour. We need only think of this set of general statements as a *theory* of the inner dynamics of human beings, as a detailed *hypothesis* concerning the determinants of human behaviour, as a theory whose credibility is a direct function of how well it allows us to explain and predict the continuing behaviour of individual human beings.[1] If its prowess in these respects proves considerable, then one has paradigmatically good reason for accepting that dynamical theory as true: for supposing that humans are indeed subject to the kinds of states posited by the theory, and that our behaviour is a function thereof as the theory describes. And one would have these reasons, note, *independently* of any appeal to the facts of one's own case. Conceivably, the facts of one's own case might even be very *different* from what the theory asserts (one is a Martian, say, with a radically alien psychology), but one could still have excellent grounds for embracing it as true of humans generally. In principle then, one's epistemic access to the minds of others need owe nothing to one's access (if any) to one's own.

Can we plausibly construe our actual knowledge of others in the fashion just outlined? Before assigning an answer to this question, let us look at the problem of other minds in broader perspective, for there is more involved here than just the question, 'How does one know there are other minds?' The facts are these: in the course of our daily affairs we display a systematic ability to explain, predict, and understand the behaviour of certain animated particulars in terms of the wants, beliefs, pains, cogitations, and other

---

[1] This fertile suggestion is Sellars'. See 'Empiricism and the Philosophy of Mind'. My debt to Sellars throughout this chapter is enormous.

psychological states and sequences to which they are presumed subject, and our facility and success in such matters is astonishing. Indeed, we are able to explain, and even predict, sequences of behaviour in others of a (kinematically described) kind novel to us, of a kind we have neither witnessed before nor undergone ourselves. The problem is to account for these abilities.

If we are going to account for them at all, I do not see how we can avoid the suggestion that we share a command or tacit understanding of a framework of abstract laws or principles concerning the dynamic relations holding between causal circumstances, psychological states, and overt behaviour. Bluntly, we share a moderately detailed general understanding or *theory* of what makes people tick. And our ability to construct explanations and predictions of their behaviour, and even of their inner goings-on, resides in our command of the general principles that constitute that theory.

Despite an initial strangeness then, the proposal that our ordinary understanding of matters psychological is constituted by a common-sense psychological theory appears to provide a unified account of all the outstanding difficulties here. The justification of that theory – call it 'the Person-theory of humans', or 'the P-theory' for short – consists in its success in accounting for human behaviour at large. Our command of that P-theory accounts for our explanatory and predictive prowess with respect to one another. And the basic justification for believing that any given particular is a person (is "another mind") is just that its continuing behaviour is most successfully explained and predicted in terms of the systematic ascription of P-states to it.

What does this P-theory look like? Recognition will be immediate, I think, for it is constituted by such familiar principles as may be found below.

(1) Persons tend to feel pain at points of recent bodily damage.
(2) Persons denied fluids tend to feel thirst.
(3) Persons who engage in vigorous activity tend to feel fatigue.
(4) Given normal attention and background conditions, persons tend to perceive the observable features (i.e. the normal colours, shapes, textures, smells, sounds, and configurations) of their immediate environment.
(5) Persons in pain tend to want to relieve that pain.
(6) Persons who feel thirst tend to desire potable fluids.

(7) Persons who are angry tend to be impatient.

(8) Persons who believe that P, where P elementarily entails that Q, tend to believe that Q.

(9) Barring preferred strategies and/or incompatible wants, persons who want that P, and believe that Q would be sufficient to bring it about that P, tend to want that Q.

(10) Persons subject to a sudden sharp pain tend to wince and/or cry out.

(11) Persons who are angry tend to frown.

(12) Persons who believe that P tend to assent to P when queried.

(13) If a person wants that P, and believes that Q would be sufficient for P, and is able to bring it about that Q, then, barring preferred strategies and/or incompatible wants, he will bring it about that Q.

This list of "P-laws" is the merest sampling of the interwoven battery commanded by any normal adult, and I state them here in but crudely basic form. These express only the gross regularities, as it were. The finer points come to be appreciated as one matures and one's experience and familiarity with human nature continues to grow (for example, one comes to learn of the relevance of analgesics, pre-empted attention, and simple shock to the rough generalization expressed in (1)). But the sample is representative, I think, and it will serve to give a rough idea of what the P-theory as a whole is like. Items (1)–(4) concern external circumstances and their inner effects. Items (5)–(9) concern various intra-mental regularities. And items (10)–(13) concern some of the inner causes of overt behaviour. The yawning familiarity of all of them bespeaks the status here being claimed for them – elements in a *common-sense* theory of the determinants of human behaviour.

We may here anticipate the objection that such "laws" as constitute the preceding list are all analytic. And being merely analytic, the objection continues, such principles cannot be supposed to constitute an empirical theory, common-sense or otherwise.

The short way with this objection is just to deny the existence of an analytic/synthetic distinction and move on. But a few remarks here may serve to surmount the objection less brusquely. Those who feel intuitions of analyticity upon viewing the likes of (1)–(13) are responding to the considerable semantic importance enjoyed by all of these sentences. They do indeed form or express a part of our collective understanding of the crucial terms involved. But a moment's reflection will reveal that these same principles also play a very active role as background "covering laws" in the common-sense explanations we regularly construct and exchange with one another, and such an explanatory role is inconsistent with their being the vacuous trivialities that the charge of analyticity implies. One can even recover or reconstruct such principles directly from an examination of our everyday explanatory practices.[2] The alleged analyticity of such principles also sits poorly with the fact that one learns to modify them, to add newly apprehended qualifications, as one's knowledge of human nature deepens. Even if there were analytic truths then, the principles at issue would not be among them.

There is one further matter that our general position here allows us to account for: the troublesome open-endedness of the behavioural "criteria" for specific P-states. Writers have often spoken of, for example, 'the behavioural criteria for pain', as if there were a finite list of types of behaviour uniquely associated with pain. But of course there is no definitive list of behaviours – not for being in pain, and even less for such P-states as 'secretly admiring the genius of Hegel'. To the contrary, there is an indefinitely large variety of kinds of overt behaviour that could, in suitable circumstances, count as behavioural evidence of your being in pain. Even your singing the *Marseillaise* could be evidence of pain, if we knew you believed that song to have analgesic properties, and could discern no other reason for your display. This complexity in the relations between inner states and their "outer criteria" is simply a consequence of the general fact that any singular hypothesis ascribing a P-state can, depending on what additional assumptions are supplied concerning one's circumstances and/or further P-states, explain any of an indefinite variety of kinds of behaviour. Accordingly, an indefinite variety of kinds of behaviour can, depending on the circumstances, count as evidence for any given P-state. In sum, the relations between inner states and overt behaviour are not one-to-one; rather, they are mediated in the complex and open-ended ways characteristic of a full-blown theory. And it is ultimately for this reason, surely, that reductive behaviourism is false. Mental states are no more reducible to overt behaviour,

---

[2] For a detailed reconstruction of an "action law" from our everyday explanatory practices, see P. M. Churchland, 'The Logical Character of Action Explanations', *The Philosophical Review*, vol. 79, no. 2 (1970).

actual and potential, than protons are reducible to vapour trails, actual and potential.

It is of course possible to be unimpressed by the P-theory. The inclination of many will be to complain about the extreme looseness of many of its alleged laws, about the vagueness of many of the terms they contain, about the dimensions of human behaviour for which it fails to provide any account, about the ramshackle nature of the theory as a whole, and so forth. While it is possible to exaggerate these shortcomings, it must be admitted that such complaints are not without considerable foundation. Indeed, I shall be urging them myself later in this essay. But in the present context such complaints are misdirected, for I am not in the least concerned to *defend* the P-theory as an accurate or adequate general account of the behavioural dynamics of human beings. The thesis of this section is only that the common-sense conception of persons is constituted, for better or for worse, by a theoretical framework of the sort outlined above. And this much, I think, is clearly correct. We have here presumed, then, only to describe common sense, not to praise it. That we might ultimately choose to bury it remains a live possibility.

13. *Self-knowledge: a preliminary look*

The aim of the preceding section was to outline a thesis, concerning the nature of our conceptual framework for persons, under which propositions about the P-states of others enjoy a non-problematic epistemic status. Of comparable importance, however, and needing explicit integration into the picture so far constructed, is the epistemic status of *first*-person ascriptions of P-states. We need to explore, that is, how one has or comes to have knowledge of one's own P-states.

It has been claimed that the construal of our ordinary psychological concepts as theoretical concepts founders on this very matter. For in one's own case, of course, one generally ascribes P-concepts to oneself non-inferentially or observationally. One is a direct spectator of one's own mental life in a way that one is not a spectator of anyone else's. How then, the objection concludes, can we represent these *introspectively* applied concepts as *theoretical* concepts?

With this objection we encounter again the same prejudice abused at length in chapter 2 – that a concept applicable in observation is *ipso facto* not a theoretical concept. We need only remind ourselves that observationality is a matter of a concept's mode of singular application, whereas theoreticity is a matter of its being semantically embedded in a framework of speculative assumptions, and the felt tension will disappear. For there is nothing inconsistent in the idea that one should be able to make reliable non-inferential applications of a concept whose semantic identity is fixed by a theory. All one needs to do is contrive a reliable habit of conceptual response to situations where the concept at issue truly applies. For as we saw in chapter 2, such is the nature of observation judgements generally. Insofar as introspective judgements are just a species of observation judgement then, there is no problem at all about the theoretical nature of the concepts they characteristically involve.

But perhaps introspective judgements are a highly special sub-class among observation judgements, a subclass possessed of unique epistemic credentials. A substantial tradition, at least, holds that they are. I have in mind here the familiar view that, at least for a certain range of P-states, introspective judgements concerning the occurrence of those P-states cannot possibly be mistaken. They are, it is variously said, incorrigible, indubitable, and infallible. If they are, then the account of them embraced in this essay must be badly mistaken, for it lacks the resources to sustain any claims of infallibility for introspective judgements. Indeed, our picture is flatly inconsistent with any such claim. The relation between any occurrence of a P-state φ and the judgement 'I am *in* P-state φ' is here held to be *causal*. It is an instance of a general habit of response to P-states of that kind. And there is nothing to preclude the possibility that in some situations that judgemental response might be prompted by some cause other than its normal antecedent – by something *similar* to φ, for example, or by something one strongly *expects* to be φ, or even by causes utterly unknown. On our picture, such judgements are no more infallible than are the deliverances of the introspective ammeter discussed at the end of §5. And there is a further respect in which the infallibility thesis conflicts with the present view. If our conceptual framework for P-states is an empirical theory, then it is possible, at the limit, that said theory be wholly false, that there are no such things as P-states, that *all* of our introspective judgements have been systematically false by reason of presupposing a false background theory. This is just conceivable, on our view, but inconceivable if introspection provides an infallible view into our own natures. If we wish to defend our more humble picture of introspection then, we must try to undermine this competing story.

The edge of the wedge is best placed, I suggest, against the following fact. Introspective judgements are clearly not infallible over the entire range of P-states. We are often mistaken in our apprehension of our own desires, fears, beliefs, emotions, and so forth. With maturity, perhaps, comes greater insight and reliability on such matters, but it is plain that infallibility characterizes at most a relatively small part of one's self-knowledge. The infallibility theorist need not deny this, of course. Infallibility was never claimed for the entire range of introspective judgements. But the difficulty begins to emerge when we ask what it is that marks off those P-states of which we have genuinely infallible introspective knowledge from those other P-states of which we have at best fallible introspective knowledge. If it is claimed that some of one's P-states are infallibly knowable, then we have a right to ask just which these are, and just what is special about them such that judgements about these, unlike judgements about the others, are actually infallible. In short, we want the "realm of the infallible" delineated in an explanatory way.

These reasonable demands have never been satisfactorily met. The range and limits of infallible introspection remain vague and disputed, and the explanations provided, of this marvellous faculty, have never got much beyond the stage of hand-waving metaphors such as "transparency to consciousness", "self-presentation", "self-validation", and the like. But metaphors like these at most mark the problem: they do nothing to solve it.

One approach alone has tried to cash out these metaphors. That approach attempts to identify the state of being in pain (for example) with the judgement *that* one is in pain. The claim is made that these are after all just one and the same thing. Once made, therefore, any judgement of that kind must be true.

This suggestion has a certain appeal, perhaps, but upon reflection it will prove fleeting. The first point to notice about the explanation proposed is that it does nothing to solve the delineation problem outlined above. Between the extremes of pain (where one's judgement is extremely reliable) and diffuse character traits (where one's judgement is highly *un*reliable), there are a great many P-states, and the spectrum on which they are severally arranged appears quite smooth. For a P-state somewhere in the middle of that spectrum, consider the feeling of hunger. Here we have a case that parallels the traditional example of pain in many respects, but our recognitional access to this feeling is fairly clearly not infallible. One can have the feeling of hunger and not realize it for some time, recognizing it suddenly in the familiar fashion ('Good lord, I'm absolutely famished!'). And one may briefly misidentify such things as thirst, or the desire for a cigarette, as hunger ('Am I hungry or not?...hmmm'). The feeling of hunger, like many other P-states, is not identical with its explicit conceptual recognition. But why the feeling of pain should be identical with its conceptual apprehension, while the feeling of hunger (for example) is not, is a matter left unexplained and unaddressed by the suggestion at issue. In this respect, at least, the suggestion appears to be imposing an arbitrary dichotomy on what is in fact a smooth continuum of cases.

But there is a more decisive reason for rejecting the conflation of the occurrence of certain P-states with one's recognition of their occurrence. Part of the appeal of this idea derives from its resemblance to the truism that there are no unfelt feelings. Let us agree that there are none. But while it is one thing to claim that all feelings are felt, it is quite another to claim that all feelings are *conceptualized* as such, are *recognized* as feelings, and as feelings of a determinate *sort*. For this latter claim is clearly false. An infant, for example, is presumably subject to a substantial range of sensations and feelings – bodily, sensory, and emotional. But being an infant he has yet to generate or acquire the conceptual framework necessary to judge *that he is thirsty*, to recognize *that he is in pain*, or to be aware *that he is having a sensation of red*. No doubt the infant feels thirst, suffers pain, and senses redly, but a judgement to any such specific effect is as yet beyond his capabilities.

In sum, the difference between having a φ-sensation, and judging that one has a φ-sensation, shows up in the following way. The making of a judgement necessarily involves the application of concepts (the concept of *pain*, for example, or *thirst*), whereas the mere having of sensations and feelings does not require the application or even the possession of any concepts at all. The sensation and the judgement are therefore two distinct things, contingently related at best.

This rejection of the identity at issue should in no way be construed as a denial of the obvious fact that infants *react* in characteristic ways to their pain, both with further P-states (distress, tension, anger) and with overt behaviour (howls, thrashing, and so on). But these reactions are all distinct from the specifically conceptual apprehension at issue, and must not be confused with it. What needs to be appreciated here is that the infant has as great a need to generate or acquire a conceptual framework with which to comprehend his own states and internal activities as he has to generate or acquire a conceptual framework with which to comprehend the states and activities of the world at large. The nascent mind must learn the parameters of its own being no less than the parameters of the universe that contains it. With time, it does learn about itself, but by a process of conceptual development that differs none from the process by which it maps the external world.

To put the matter thus is to adopt a very different approach to the philosophy of mind than is embraced by the bulk of the modern tradition. For we are here denying the near-universal conviction that the mind or self is somehow "better known" to itself than is the universe around it. Among philosophers of the last three or four centuries, Immanuel Kant stands almost wholly alone in insisting that knowledge of oneself is entirely on a par with knowledge of the world external to oneself.[3] Moreover, to put the matter thus (and here we part company with Kant) is to highlight the possibility of conceptual *progress* in the matter of self-comprehension, and to raise the question of the adequacy of our current framework for inner states. That is to say, behind the question of what local perils threaten the propriety of this or that first-person ascription of this or that P-predicate, there is the larger question of the propriety of the relevant background framework – the P-theory – itself. On the view here embraced it is conceivable, first, that said framework is inadequate as a representation of our internal reality, perhaps profoundly inadequate, and second, that one might learn to comprehend and report one's internal states and activities within a different and more adequate framework. We have seen a sketch (in §4) of a general transformation in the conceptual matrix of non-inferential judgement concerning the external world. The suggestion here is that a general transformation in the conceptual

---

[3] Against the idealists, for example, he remarks, 'What they have failed, however, to recognize is that both [the objects of inner sense and the objects of outer sense] are in the same position; in neither case can their reality as representations be questioned, and in both cases they belong only to appearance . . .', *Critique of Pure Reason*, B 55.

matrix of non-inferential judgement concerning one's internal states would follow the very same lines.

Before examining this idea in more detail, however, let us take a more penetrating look at the framework we have just proposed to displace. For the P-theory is in fact a marvellous intellectual achievement. It gives its possessor an explicit and systematic insight into the behaviour, verbal and otherwise, of some of the most complex agents in the environment, and its overall prowess in that respect remains unsurpassed by anything else our considerable theoretical efforts have produced. Additionally, it embodies certain curious features not found in theories generally, features central to its success, on the one hand, but problematic for its unification with the rest of science on the other. Questions of relative adequacy, of possible reduction, or of potential displacement, therefore, cannot adequately be discussed without a detailed appreciation of the special nature of the P-theory, and it is to this matter that we now turn.

14. *The incongruent nature of the P-theory*

Among P-predicates generally, a certain subset merits special attention. Consider the following examples.

. . . believes that p	. . . infers that p
. . . knows that p	. . . hopes that p
. . . desires that p	. . . is dismayed that p
. . . fears that p	. . . intends that p
. . . sees that p	. . . prefers that p
. . . suspects that p	. . . is thinking that p

This list could be lengthened considerably, but these examples will serve to illustrate the class of P-predicates I wish to discuss. Expressions of this sort have long intrigued philosophers of mind and logicians alike. In the first instance, such expressions are clearly central to our conception of ourselves as persons – as perceivers, thinkers, deliberators, and so forth. And in the second instance, the logical form of such expressions has always been cause for special attention since certain elementary logical operations (such as substitution of coextensive expressions), thought to be universally permissible on sentences at large, fail to preserve truth when applied to sentences nested in place of 'p'. Conjointly, these facts have suggested to many that in the special logical properties of the predicates at issue we have a specific reflection of the unique and non-material nature of the mental. That conclusion is no doubt seriously overdrawn, but it remains for us to penetrate the exact significance of these unusual expressions.

In hopes that the exercise will ultimately throw light on the expressions listed above, let us shift our attention and examine a second list, a list whose elements display a superficially similar pattern.

> . . . has a length in metres of n
> . . . has a volume in cubic metres of n
> . . . has a velocity in metres per second of n
> . . . has a mass in kilograms of n
> . . . suffers a force in newtons of n
> . . . has a temperature in kelvins of n
> . . . has a charge in coulombs of n
> . . . has an energy in joules of n
> . . . has a magnetic flux in webers of n

This list, if completed, would be even longer than the first, and we have in its elements the characteristic expressions of what might loosely be called mathematical science. How should we construe the elements of this second list? As I see it, we are looking at a list of predicate-forming expressions. Each of the above expressions yields an entire family of distinct predicates – determinates under each determinable – as we substitute for 'n' distinct singular terms for distinct numbers. That is, I propose to construe these expressions as having the form $\varphi(n)\ldots$, as predicate-forming functors taking singular terms for numbers in the argument place held by 'n'.

There are, to be sure, other ways of construing them. One might interpret 'a has a length of 3 m' as having the form $aL3$, as asserting a relation between a and the number 3. Alternatively, one might recast the example as 'the length in metres of a is 3', represent its logical form as $l(a) = 3$, and see singular-term-forming functors in such examples, rather than predicate-forming functors. This third alternative will generally be the most convenient, by virtue of the economy it permits in the expression of familiar scientific laws. That economy results from the uniqueness implication built into the use of definite singular terms rather than predicates. On the other hand, the third alternative is less general for that very reason, being incapable of handling cases where the uniqueness implication is false or inappropriate. The first two alternatives have no difficulty in this regard, and can incorporate a uniqueness condition whenever needed. (See, for example, the rendition of Newton's second law in the text below.) The remaining issue between the relational and the predicative alternatives may be passed over for the time being, for the parallels I wish to draw will emerge with equal clarity on either construal. Just the same, I shall here stick with the initial (predicative) construal, for reasons that will become clear presently.

What must be appreciated about predicate-forming expressions of the kind at issue is the enormous extent to which they amplify our conceptual resources. With them, we can conceive and express *empirical properties, the relations between which are the relations between numbers.* We can both conceive, and find instanced in the warp and woof of empirical reality, relational configurations to which we were blind prior to the introduction of conceptual machinery of this novel kind. Many of these relations, of course, will be comparatively uninteresting: the mass of my car in kilograms is one thousand times my height in metres, for example. There is nothing of any moment here. But others may be of great interest, being instances of relations that hold *universally.* And if they are, we can give expression to those general patterns, using expressions of the kind at issue and the quantificational resources already in use, as in the following characteristic example.

> For any body *x*, there is a unique *f*, *m*, and *a* such that *x* suffers a net force of *f*, *x* has a mass of *m*, *x* has an acceleration of *a*, and *f* = *m* × *a*.

Canonically,

$$(x)\,[Bx \supset (\exists f)\,(\exists m)\,(\exists a)$$
$$[F(f)x \,\&\, (\imath)\,(F(\imath)x \supset f = \imath) \,\&\,$$
$$M(m)x \,\&\, (\imath)\,(M(\imath)x \supset m = \imath) \,\&\,$$
$$A(a)x \,\&\, (\imath)\,(A(\imath)x \supset a = \imath) \,\&\, (f = m \times a)]].$$

With the advent of predicate-forming expressions such as these, the infinite ranks of systematic relations holding in the domain of numbers are all made available for duty in the service of overtly empirical theory. We gain a mode of description more nearly equal to the relational intricacies of the empirical world. Pythagoras, in his conviction that the relations between all things are ultimately the relations between numbers, appears to have anticipated the representational power and potential of theories using expressions of the kind at issue. Even so, he would surely be struck speechless by what we have since managed to achieve through their use.

While we must concede its fertility, however, the domain of numbers is not the only domain of abstract entities in which interesting and exploitable relations hold. Is it perhaps possible that there could be non-Pythagorean theories, theories exploiting the relations holding in other abstract domains? Theories containing predicate-forming expressions taking as arguments singular terms for abstract entities other than numbers? Abstract entities like *propositions*, perhaps? Certainly it is at least conceivable that a theory could be of this kind. And in looking back at the list of psychological expressions with which we began this section, expressions for which we have already found independent reason to count as elements of a theory, it rather looks like we may have just such an animal.

By way of exploring this suggestion, let us construe the likes of '... believes that *p*' as predicate-forming functors taking as arguments singular terms for propositions. We need at this stage commit ourselves to no special view as to what propositions *are*. If we wish to construe 'that', in the relevant contexts, as a mere quoting device, then the singular terms are just quoted sentences, and propositions need be nothing more offensive than sentences. Construe 'that' differently (as is highly advisable), and we can construe propositions differently. The thesis here being suggested is so far neutral in this regard. What is important is that with the advent of predicate-forming expressions such as those on our first list, we can conceive and express *empirical states, the relations between which are the relations between propositions.*

To complete the proposed analogy with the structure and intellectual strategy of Pythagorean theories, we need no more than a positive answer to the following question. Does our common-sense understanding of the general *empirical* relations holding among and between the various propositional attitudes to which one is subject depend on or reflect the various *logical* relations (equivalence, entailment, inconsistency, and so forth) holding among the propositions in those attitudes? Well of course it does. The following examples approximate, I think, some of these principles embedded in common sense, and will illustrate the parallel with the intellectual strategy of what I have called Pythagorean theories. Expressions of the form '$\varphi(p)x$' are to be read 'x φ's that p'.

$(x)\,(p)\,[(\text{Wants}(p)\,x\;\&\;\text{Discovers}(p)\,x) \supset \text{Pleased}(p)\,x]$

$(x)\,(p)\,[\text{Fears}(p)\,x \supset \text{Wants}\,(\sim p)\,x]$

$(x)\,(p)\,(q)\,[(\text{Believes}(p)\,x\;\&\;\text{Believes}(\text{if }p\text{ then }q)\,x) \supset$
    $(\text{Believes}(q)\,x \lor \text{Reconsiders}(p)\,x \lor \text{Reconsiders}(\text{if }p\text{ then }$
    $q)\,x)]$

$(x)\,(p)\,(q)\,[(\text{Wants}(p)\,x\;\&\;\text{Believes}(\text{if }q\text{ then }p)\,x) \supset (\text{Wants}(q)\,x$
    $\lor\;(\exists s)\,(\text{Wants}(s)\,x\;\&\;\text{Believes}(\text{if }q\text{ then }\sim s)\,x))]$[4]

We do indeed understand the relevant empirical regularities, such as they are, in terms of the logical relations they embody or reflect. A moment's perusal of examples of the reader's own construction will corroborate the claim that these principles support predictions, subjunctive conditionals, and explanations of precisely the kinds proposed and accepted in everyday social commerce. As baldly presented above, no doubt they want qualification in certain respects, but those details are of no particular concern to us here.

I conclude then that the common-sense conception of persons is a non-Pythagorean theory whose central expressions are predicate-forming functors taking singular terms for propositions as arguments, and whose rough principles (or a central core of them) exploit relations in the domain of propositions rather than relations in the domain of numbers.

---

[4] A point concerning notation: it must be kept in mind here that what substitute for the variables 'p' and 'q' in the nested contexts are not sentences but singular terms *formed* from sentences. In English, the job of singular-term formation is done by affixing 'that' to the sentence in question. Here, we can stipulate that the special context of the relevant argument place does that job. Treating those contexts as occupied by singular terms raises a further complication when we wish to form compound singular terms such as ⌐if p then q⌐, since 'if-then' is properly a *sentence* operator rather than a singular-term operator, and so for the other standard connectives as well. We must stipulate, therefore, that in such contexts 'if-then', for example, forms singular terms from other singular terms, and that ⌐if p then q⌐ in such contexts is a singular term naming the proposition expressed by the normal English conditional formed with the sentences used to form the singular terms ⌐p⌐ and ⌐q⌐ respectively as antecedent and consequent respectively. If such restrictions are observed, then we may quantify into the relevant places in good *notational* conscience at least. The model to follow here is just elementary number theory, in which, for example, we can form singular terms such as ⌐$x+y$⌐ from the singular terms ⌐x⌐ and ⌐y⌐.

Having blocked out a rough position here, let me now try to fill in some missing details and sharpen some important contrasts. The first matter demanding our attention is the residual issue between the predicative and the relational construal of contexts like '... believes that *p*'. The story to this point may be welcomed by those of a Platonistic inclination: not only does my reconstruction of the P-theory involve wholesale quantification over propositions; it may also encourage the Platonistic view that having a belief that *p*, to take just one intentional state, is ultimately a matter of the *self* standing in some suitable *relation* to an abstract entity – the *proposition* that *p*. For if we adopt the relational instead of the predicative construal embraced earlier, that is precisely what we shall have.

That line of thought, it seems to me, misrepresents the nature of the P-theory. The idea that believing that *p* is a matter of standing in some appropriate relation to an abstract entity (the proposition that *p*) seems to me to have nothing more to recommend it than would the parallel suggestion that weighing 5 kg is at bottom a matter of standing in some suitable relation to an abstract entity (the number 5). For contexts of this latter kind, at least, the relational construal is highly procrustean. Contexts like

> *x* weighs 5 kg
> *x* moves at 5 m/s
> *x* radiates at 5 J/s

are more plausibly catalogued with contexts like

> *x* weighs very little
> *x* moves quickly
> *x* radiates copiously.

In the latter three cases, what follows the main verb has a transparently *adverbial* function. The same adverbial function, I suggest, is being performed in the former cases as well. The only difference is that using singular terms for numbers in the adverbial position provides a more precise, systematic, and useful way of modifying the main verb, especially when said position is open to quantification.

These considerations lead us straight back to the predicate-forming functor construal of the expressions at issue, for the relevant argument place in such functors is essentially an adverbial position. In the case of Pythagorean theories, therefore, we are not bound to a relational interpretation of their characteristic expressions. On the contrary, the situation is most naturally captured by an adverbial interpretation. And precisely the same is true, I suggest, of the characteristic expressions of the P-theory. In contexts like '... believes that Tom is tall', the clause 'that Tom is tall' – a singular term for a proposition – serves an adverbial function as an element of a complex predicate.

Concerning the quantification over propositions displayed in the illustrative examples of P-theoretic laws, my own conscience is clear. For I am not here recommending either the P-theory or the practice of quantifying over propositions. My thesis is only that the P-theory is most accurately construed as a theoretical framework involving systematic quantification over propositions. I am quite prepared to consider the possibility that the P-theory is intellectually bankrupt for that very reason, propositions being illegitimate entities not fit for quantification.

As it happens, however, I doubt the P-theory is bankrupt for quite this reason. Given a theory of meaning along the lines sketched in chapter 3 – a theory that supplies a notion of sameness of meaning – we can no doubt contrive a legitimate place in our ontology for something answering to propositions. And in any case, the P-theory is no less a workable affair even if we construe propositions as English *sentences*. The quantification in question is therefore legitimate, at least in principle. This is not to deny that the P-theory has serious problems, nor that those problems are deeply involved with the matter of propositions. But it is to reject the specific complaint registered above.

Let me stand back and summarize. I have two purposes in drawing attention to the structural features of our common-sense conception of ourselves outlined in the preceding pages. The first is to highlight the similarities between the structure of our self-conception and the structure of certain paradigmatic theories. It is to illustrate from yet another angle the theoretical (systematic and speculative) nature of the common-sense conceptual framework for persons. But second, and more important, it is to isolate the crucial difference between our conceptual framework for persons and our growing conceptual framework for the rest of reality. The P-theory is unique in that it uses complex predicates ascribing *propositional* rather than numerical "attitudes", and its general principles exploit the *logical* relations holding among the propositions in those attitudes. And it is primarily for this reason, I suggest, that our self-conception stands in such stubborn, disconnected isolation from our growing conception of the rest of nature. For that grander conception is Pythagorean to the core, and it is *prima facie* problematic how the incongruent likes of the P-theory can be made a conceptually continuous part of it.

15. *The mind/body problem*

If the preceding sections are even roughly correct, then the traditional mind/body problem assumes a highly specific form. The question of the relation between the familiar psychological states on the one hand, and the neurophysiological states of the body and central nervous system on the other, becomes a question of the relation between the ontology of one theory – the culturally entrenched P-theory – and the ontology of another theory – the general theory of human neurophysiological activity that empirical psychology is in the painful process of constructing. Our problem here is just another case of one theory threatening another *theory* with either subsumption or displacement. And if this is the form the issue takes, then the range of possible outcomes is finite, and our own intellectual history in other areas provides a rich store of possible parallels from which to draw guidance. I shall now attempt to canvass the relevant possibilities.

(A) To begin with, let us consider the several dualistic alternatives in which the P-theory is neither reducible to nor replaceable by any materialistic theory comprehending the structure and activity of the human nervous system. We must here consider three possibilities.

(i) The P-theory may be irreducible and irreplaceable because it is already an essentially adequate general description of an autonomous, non-material dimension of reality – the realm of "intentional phenomena", perhaps. A rough historical parallel here would be the clear emergence, by 1910, of Maxwell's electromagnetic theory as an irreducible account of a non-*mechanical* dimension of physical reality. Many had thought that electromagnetic phenomena would prove to be understandable in terms of elastic deformations and oscillations in the subtle substance of an universal aether, that the laws of electromagnetics would turn out to be just highly special instances of or consequences of familiar mechanical laws.

Most impressively, it proved not to be so. The notion of electric charge, therefore, assumed its proper place alongside the notion of mass (and length and duration) as one of the fundamental parameters of reality. Likewise, it is conceivable that the P-theory will prove irreplaceable and irreducible as well, that psychological phenomena will prove to be a similarly autonomous and fundamental aspect of reality. We might call this possibility 'simple dualism'.

(ii) A different (and *a priori* more likely) possibility is that the ontology of the P-theory will reduce alright, but to a more general domain of phenomena that is itself autonomous and non-material, a domain of phenomena of which the specifically human psychology reflects but a small part. In other words, the P-theory, though failing to find a materialistic reduction, may prove reducible to a general science of "ectoplasmic essences", say, or to some other non-materialistic theory. (An historical parallel here would be the failure of the many attempts (e.g. Huygens', Newton's) to give geometrical optics a *mechanical* reduction, and its final reduction to electromagnetic theory.) We might call this second possibility 'reductive dualism'.

(iii) The third possibility here under assumption (A) is one that to my knowledge has never been cited before, but it is real just the same. Specifically, the P-theory might prove to be replaceable by some more general theory of "ectoplasmic essences", say, as in (ii), but to be irreducible to that more general theory. The ontology of the P-theory would thus be eliminated in favour of the ontology of the more general theory that displaced it. We might call this possibility 'eliminative dualism'! It is perhaps not surprising that this possibility has gone unremarked, since the preservation of the common-sense ontology of the mind has always been part of the dualist's sales-pitch. Let it be noted then that the demise of our common-sense P-theoretic ontology is every bit as possible in a non-materialistic ontology as it is in a materialistic ontology.

(B) If dualism is not essentially hospitable to our common-sense psychological ontology, neither is materialism essentially hostile to it, for it may turn out that the P-theory is both replaceable by and reducible to an adequate theory of the central nervous system. That is to say, a thorough understanding of our neurophysiological structure and activity may provide predictive and explanatory resources that

(1) render the P-theory superfluous so far as accounting for human behaviour is concerned, and

(2) entail, or contain as a substructure, a clear and equipotent image of the P-theory.

It is this expectation, more or less, that constitutes the identity theory (reductive materialism). Thoughts, pains, and desires, it is anticipated, will turn out to be identical with special states or events of the central nervous system, in just the fashion, it is said, that the temperature of any body turned out to be identical with the mean kinetic energy of its constituent molecules. Our familiar categories will thus survive, by being translated into a broader ontological setting.

No doubt the identity theorist's optimism on point (1) is well founded. In many respects the current theoretical situation concerning mental phenomena must be reckoned the contemporary analogue of the theoretical situation concerning "vital" phenomena as it existed in the last century. There, it will be recalled, the phenomena surrounding living tissue remained to find their proper place in the rapidly expanding scheme of things that chemistry and thermodynamics were constructing. Many thought that living tissue contained or instanced a "vital principle" that was entirely distinct from anything that mere matter could display, however organized it might be in its merely physical respects. For there was a gulf apparently fixed between living and non-living matter: the fact that inert matter could always be recovered from living tissue, but never, apparently, the other way around; the fact that living tissue reproduced itself; the perplexing facts of metabolic activity and the unparalleled mystery of autonomous morphological articulation; all of these things conspired to suggest a novel and autonomous agency at work.

The strength of the suggestion, of course, was inversely proportional to one's familiarity with the great complexity of actual and possible chemical phenomena, and with the explanatory potential of a dynamical and structural chemistry as it was taking shape in the last century. For it was plain that living tissue, after all, had mass, transformed chemical energy into kinetic and thermal energy, and decomposed without obvious remainder into wholly familiar elements. Chemical theory already contained both the conceptual and the technical resources for a systematic attack on the problem of living tissue, construing it as a question of the chemical/structural/ dynamical organization of matter. By contrast, the term 'vitalism' fronted an empty shell; it named no significant body of theory, no programme of research, no promise of convergence with gains already won. It was negligibly more than the denial of the materialism it opposed. On balance then, materialism held out vastly the better hope for an adequate account of biological phenomena, and the optimism of its adherents was justified quite independently of the almost uninterrupted stream of successes that subsequent effort brought to it.

Returning now to the sentient/intelligent behaviour of humans, the prospects for a materialist account must be reckoned in much the same fashion. The background weight of a materialist ontology is orders of magnitude greater now than ever before. The potential it provides for explaining the behavioural properties of humans in terms of the material organization of the nervous system is enormous, immediately available, and partially realized already. And like vitalism, dualism fails to provide any competing conceptual resources, being negligibly more than the denial of the materialism it opposes. And there is a further point: we must reckon also the continuity that nature displays between our own case and that of creatures with little or no sentient/intelligent capacities. The gulf that appeared to divide the living from the non-living finds no parallel in the case of purposive or intelligent behaviour. The continuity with the rest of the animal kingdom is there to behold. And the continuity is vertical as well as horizontal, for we must not forget our own evolutionary history, and its wholly material beginnings.

So far point (1). But point (1) is not enough. Essential also to the identity theorist's position is the claim of point (2) that the P-theory will *reduce* to its materialistic successor. And here it is rather less clear that optimism is justified than most identity theorists appreciate. Criticism of this second claim arrives from two quite different directions, representing the two major materialist alternatives to the identity theory. It is to these that we now turn.

(C) The less radical of the two criticisms stems from a popular view closely related to the identity theory, but distinguished sharply from that view by its equally materialist adherents. I am here referring to Hilary Putnam's functionalism.[5] According to

---

[5] See Hilary Putnam, 'Minds and Machines', in *Dimensions of Mind*, ed. Hook (New York, 1960); also 'Robots: Machines or Artificially Created Life?', *Journal of*

this view, psychological states are functional states in the sense that for any being to have a psychology (to be the subject of psychological states) is for it to instance or embody a certain functional organization among its sensory inputs, internal states, and motor outputs. Talk of psychological states is therefore ontologically neutral, holds Putnam, since descriptions at that level are innocent of any commitments as to the nature or constitution of whatever it is that instantiates the relevant functional organization. And this is as it should be, it is claimed, since it is clearly possible that the same functional organization be realized, embodied, or instantiated in creatures or systems of widely different constitutions (e.g. cellular, crystalline, gaseous, etc.). Accordingly, psychological descriptions are not reducible to descriptions concerning any of the various substances that might instantiate them. They are descriptions at a level abstracted from such matters.

Thus emerges the essential point of difference with the identity theory: it is the reducibility of psychological descriptions that is denied by Putnam. And the reason for the denial is the multiplicity of different substrata that can instantiate those descriptions. Any attempt to reduce psychological descriptions to the specific categories characteristic of any one such substratum would be inconsistent with their applicability to the remaining kinds of substrata. Our embarrassment, ironically, is one of riches. We might attempt to effect a "disjunctive reduction", but this notion makes little clear sense, and the open-ended nature of the disjunction(s) would remain a serious difficulty even if it did.

These considerations are compelling. Held out to us is the possibility of a consistent materialism with neither the burden of defending psycho-physical identities, nor the burden of denying the mental altogether. And the premiss that inspires all this is no doubt true: beings of an alien physical makeup might well be persons (i.e. constitute models for the P-theory) no less than ourselves. Even so, I think Putnam's position is importantly mistaken, and I shall now try to explain why.

In the first place, in arguing that psychological descriptions are purely functional descriptions, it is apparently essential to Putnam's position to deny that the collected lore of our common-sense psychology constitutes a corrigible empirical theory. But on the strength of the many considerations adduced in the present essay, I submit that it is quite clear that it does enjoy that status. The only objections I am aware of authored by Putnam (in 'Robots: Machines or Artificially Created Life?') are easily handled. The first takes the form of a *reductio* and points out that on the view at issue the existence of the familiar mental states is made contingent on the empirical virtues of the common-sense psychology that postulates them. This *reductio* is disarmed by embracing its wholly correct conclusion. The second objection presumes to find an inconsistency between the theoretical nature of P-predicates and their frequent use in first-person observation reports. But as we saw in chapter 2, concerning observation predicates in general, these two features are wholly compatible. In sum, there is nothing here to undermine the view that the P-theory is a corrigible empirical theory.

And in the second place, the multiple instantiation premiss conceded earlier does not entail the irreducibility of psychological descriptions. Even if full-fledged models for the P-theory can indeed be realized in a variety of different material substrata (as is likely), it does not follow that the P-theory is materialistically irreducible. *For there may yet be a characterization, of all and only such models, in physical terms of a kind sufficiently fundamental to encompass all of them,* a characterization to which, as it happens, the P-theory reduces. It is entirely possible, for example, that we, the gaseous Nebularians, the crystalline Plutonians, and any other persons lying about are all 'super-heterodyning negentropy flowers', where this expression is a part of the vocabulary of some future and more fully articulated version of statistical thermodynamics, a theory of awesome generality as it stands. And if there is a univocal characterization of all of us in terms of such a theory, it is of course possible that the P-theory prove reducible within it. For my own part, and for reasons yet to be outlined, I would not reckon this second probability as very high; but the possibility is all that the present context requires. Contrary to Putnam, our common-sense psychological descriptions are not of an essentially irreducible kind.

On the other hand, they may well turn out to be irreducible in fact. Putnam's view may succeed as prophecy where it failed as analysis, and the possibility it represents is important. I shall try to explain. First, suppose that future research confirms our expectation that the P-theory is instantiable in a variety of material substrata. Second, suppose that in each such case we can explain how the functional organization at issue (that implied by the P-theory) arises from the physical makeup peculiar to that substratum. Materialism, that is, turns out to be sustained. But suppose, finally, that despite all this detailed insight, we are unable to fund, in purely physical terms, a *unitary* characterization of all and only those particulars that are models for the P-theory. In that event the P-theory would have to be counted irreducible. Accordingly, it would have to be seen as providing at most an abstract functional characterization, *since in that event persons turn out not to constitute a natural kind.* The discoveries here envisaged would have the effect of "denaturing" the P-theory: its many principles could no longer be viewed as law-like generalizations concerning a certain natural kind, but only as stipulative characterizations of a certain non-natural kind. (Compare, for example, the set of abstract principles specifying the operations of anything that is to count as an *arithmetical calculator.*) Putnam, in short, would turn out to have been right. For this sort of outcome I can cite no clear historical example, but had simple pocket calculators always grown on trees our history might well have contained one.

So far the sympathetic prognosis. Unfortunately, this prognosis is, like the identity theory (and the first two forms of dualism), blissfully optimistic in one highly problematic respect. To put it bluntly, it assumes that *we* are genuinely models for the P-theory; it assumes that the categories of the P-theory have sufficient integrity to survive intact the light that future theory and experiment is going to shed on the operations of the central nervous system. Even as a functional characterization of ourselves, that is, the P-theory may turn out to be taxonomically cockeyed, radically incomplete, and altogether too confused to merit continued use, when compared with the much superior functional characterizations that an adequate theory of the central nervous system (CNS) can be expected to provide. Whether or not the P-theory will survive then, even as a purely functional characterization, will depend on how good and how useful a characterization it turns out to be relative to the alternative modes of self-understanding that future research will surely produce. A live possibility is that it will prove a comparative cripple, and this suggestion will serve to introduce the fourth major position here.

(D) The last of the materialist views is the least óptimistic about the ultimate fate of the P-theory. Where the dualist anticipates some non-materialist triumph for the P-theory, the identity theorist anticipates its survival by materialist reduction, and the functionalist expects its survival in denatured form, a fourth view foresees its complete demise. The eliminative materialist holds that the P-theory, not to put too fine an edge on the matter, is a *false* theory. Accordingly, when we finally manage to construct an adequate theory of our neurophysiological activity, that theory will simply displace its primitive precursor. The P-theory will be eliminated, as false theories are, and the familiar ontology of common-sense mental states will go the way of the Stoic pneumata, the alchemical essences, phlogiston, caloric, and the luminiferous aether.

Why take such a view seriously? To see why we should, let us put the onus on the opposition for a moment and ask why the alternatives (A)–(C) discussed above are so uniformly optimistic about the P-theory's survival. That optimism derives, I suggest, from the uncritical assumption that the P-theory is a *true* account of the dynamics of human behaviour, that the ontology of the P-theory is (non-fundamental, perhaps, but) *real.* False theories can be displaced, it will be said, but a true theory need fear no threat from further research.

Such confidence in the categorial framework of the P-theory is not difficult to understand, historically. That said framework is a theoretical one is a relatively recent insight, and without this insight that framework is likely to appear, as it has appeared, as something *manifest* rather than as something conjectural. But while this may explain some of our confidence, it does nothing to justify it. There is only one place to look for justification, and a measure can indeed be found in the actual success of the P-theory as the vehicle of our mutual understanding. But it is just here that the case is so problematic, if one looks at all closely, for its success is in many respects radically incomplete. Its comprehension both of practical and of factual reasoning is sketchy at best; the kinematics and dynamics of emotions it provides is vague and superficial; the vicissitudes of perception and perceptual illusion are, in its terms, largely mysterious; its comprehension of the learning process is extraordinarily thin; and its grasp of the nature and causes of mental illness is almost nil. In sum, there is much about ourselves, *qua* perceiving/reasoning/passionate creatures, that we do not understand at all, and much about our behaviour that we cannot begin to explain. The P-theory gives us what is obviously a superficial gloss on a very complex set of phenomena. Accordingly, its actual success is a thin thread on which to hang much faith in its ultimate fate, and its extensive impotence and plain superficiality provide a rich medium in which the seeds of doubt can grow. Moreover, the P-theory has been in active and continuous service in all human cultures for as long as history records, and despite the obvious room for improvement, the P-theory has undergone no significant development or improvement in all those thousands of years. Viewed in this light, the P-theory appears as a dead-end approach to the problem of human nature, an approach that has survived only because of practical inertia and a dearth of developed alternatives. Its infertility is readily ascribed to the poverty of its basic categories, and it is not unreasonable to expect that they, and the common-sense laws that comprehend them, will prove too confused and superficial to find any natural reconception or reform-ulation within the framework of a truly adequate theory of our internal activity.

Reducibility is a matter of degree, of course, and it is unlikely that the P-theory will turn out to bear *no* relation to the true account of things, but this is easily conceded by the eliminative materialist. His point is only that the empirical virtues of the P-theory are sufficiently meagre that it is unreasonable to expect that it will reduce with sufficient smoothness to float an ontological reduction, and more reasonable to expect that it will simply be dropped, forsaken, as it were, for a prettier face. The ease with which this might happen is no doubt difficult to imagine while that prettier face remains hidden from us still, but the lessons of our own intellectual history should free us from the narrow-minded prejudice this difficulty represents.

If the P-theory is indeed a theory, then the long-term outcome just described must be reckoned a real possibility. And it will do no good to protest that introspection reveals directly the existence of thoughts, desires, passions, and sensations. For introspection, as we have seen, consists in the exercise of those acquired habits of conceptual apprehension where the (current) medium of appre-hension is the P-theory itself. As well insist on the reality of caloric fluid – having been trained to conceptualize the relevant experience in terms of that theory – on grounds that its existence is perceptually obvious. Eliminative materialism, therefore, is very much a live option.

This completes my general outline of the possible futures the P-theory may discover, and of the nature of the issues between their several proponents. How should we estimate the relative prospects for each of them? One important point here obtrudes itself immediately. Given the historical/theoretical/empirical background surveyed on pp. 109–10, the prospects for dualism must be logged as negligible. The issue shapes up as a choice between (B), (C), and (D). Here the matter is more difficult to decide, and must surely prove impossible to settle decisively before actually completing the materialistic account of human behaviour and internal activity which the remaining parties to the debate expect. For the remaining issues all depend crucially on the scope of that theory and on the relation it bears (or fails to bear) to the P-theory, and both these matters depend crucially on the actual content of that yet-to-be-invented theory.

Short of precognition then, the only relevant premises available to us concern the discernible virtues and shortcomings of the P-theory. As I judge its shortcomings to be very substantial indeed, I confess a strong inclination towards eliminative materialism, but I shall make no stronger claim at this point. The poverty of the P-theory becomes the subject of discussion again in the final chapter.

16. *The expansion of introspective consciousness*

At the close of §5, non-inferential judgements concerning one's own psychological states were cast as epistemologically undistinguished members of the class of observation judgements generally, and that suggestion was more fully explained and defended in §13. The central theme of the relevant discussion to this point is that self-perception consists in the disposition-governed occurrence of conceptual responses to one's internal states, responses made within whatever matrix of self-understanding one has developed or acquired. Our current matrix of response is the P-theory, but we may safely presume that this has not always been the case. How our primitive (pre-linguistic) ancestors thought of themselves we may never know, but our conceptual framework for our own psychology has certainly undergone enormous development in the interim, evolving in step with the evolution of our brain capacity, linguistic skills, and social integration. The "emergence of self consciousness", a phenomenon thought by some to pose a unique difficulty for any evolutionary theory, is therefore nothing essentially mysterious. In the context of the view being proposed, it appears as just a special case of learning – a case of learning to perceive and understand *oneself.*[6] If one can learn to make systematic responses to states of the environment, it is no mystery that one can learn to make systematic responses to states of oneself (one is already doing the latter when one does the former). And, as with learning in general in a social and linguistic context, it can become a cultural snowball. A shared framework, once in action in many individuals, evolves and grows in response to the further learning it has made possible. The socially supplied machinery for self understanding eventually permits a depth and detail of introspective awareness that would be profoundly unlikely in any individual raised without the benefit of that cultural heritage. It is an important truth then, if a somewhat ironic one, that the high degree of *self*-awareness one currently enjoys is ultimately a social phenomenon in its causal history.

   If we humans have come this far, must we end the journey here? Is the P-theory the complete and final word on the inner nature of Man? Must our introspective vision be forever limited to the particular categories that theory provides? These are rhetorical questions, for the answers, surely, are uniformly in the negative. On all three of the materialist alternatives discussed in the last section (and on two of the three dualist alternatives as well), there is a great deal about our psychology that remains to be learned. Consequently, there is much scope for possible expansion and/or transformation in the form and substance of our introspective activities as well. In fact, there is every bit as much room for expansion with respect to the apprehension of one's internal states as we found there to be with respect to our apprehension of the external world, and for the very same reasons. The framework within which our non-inferential conceptual responses are currently made is less than wholly adequate to reality (according to (B), it is insufficiently general; according to (C), it is merely a functional description; and according to (D), it is sufficiently confused to be dismissed as false.) It is therefore possible that such responses

   [6] What remains mysterious, of course, is the general phenomenon of *learning.*

might come to be made within the framework of a more penetrating theory of our internal states and activities. And, as before,
the difference between the old inner vision and the new may be
substantial.

In detailing this sort of outcome back in §4, I had an important
advantage: concerning the world at large, we already possess
a theoretical network whose collected virtues far surpass those of
the common-sense scheme. The "new" scheme being available,
one can sketch at one's leisure case after case of its induction into
an observational role; one can begin to gain an active feel for the
new mode of perception as a closely integrated whole; and one can
appreciate in detail how the density of apprehended information
would go up sharply. In the present case, unfortunately, the path
is not so smooth. The anticipated super-framework suitable for
reducing the P-theory (or for explicating its instantiation in us, or
for displacing it entirely) is not yet within our grasp, nor can we
claim to discern its outlines with any notable clarity. A guided
tour of its projected non-inferential applications is therefore out
of the question.

Even so, it is easy to look ahead. While the P-theory is in no
immediate danger of replacement, especially in regard to its conceptual machinery for propositional attitudes, our present knowledge of the behavioural dynamics and inner activities of *homo
sapiens* extends far beyond what that theory supplies. And for many
of these inner states, it is already plain that they can be brought
within the scope of non-inferential conceptual apprehension.

Consider, to begin with, the possible reconception – in non-
P-theoretic terms – of some of the familiar states and processes
already apprehended by standardly-acculturated introspection.
(Whether this reconception takes place within the context of a
type/type reduction, a neurological explication of molar functional
organization, or a wholesale conceptual displacement, I here leave
open. The important point here is as follows: if we are willing to
count our inner states as introspectable under descriptions provided
by the P-theory, then we must also be willing to count them as
introspectable (in principle) under the descriptions provided by
whatever theory promises to subsume or displace the P-theory.)
Such reconceptions may correspond closely or only distantly to
our current conceptions, but possible examples are not hard to
find. The considerable variety of states currently apprehended in

a lump under 'pain', for example, can be more discriminately recognized as sundry modes of stimulation in our A-delta fibres and/or C-fibres (peripherally), or in our thalamus and/or reticular formation (centrally). What are commonly grasped as "after images" can be more penetratingly grasped as differentially fatigued areas in the retina's photochemical grid, and the chemical behaviour of such areas over time – specifically, their resynthesis of rhodopsin (black/white) and the iodopsins (sundry colours) – is readily followed by suitably informed introspection. The familiar "phosphenes" can be recognized as spontaneous electrical activity in the visual nervous system. Sensations of acceleration, and of falling, are better grasped as deformations and relaxations of one's vestibular maculae, the tiny jello-like linear accelerometers in the vestibular system. Rotational "dizziness" is more perspicuously introspected as a residual circulation of the inertial fluid in the semicircular canals of the inner ear, and the increase and decrease of that relative motion is readily monitored. The familiar "pins and needles" at a given site is more usefully apprehended as oxygen deprivation of the nerve endings there located. And so forth. What all of these shifts amount to, of course, is nothing more than a recalibration of one's self-monitoring mechanisms: a shift in the interpretation functions that govern the content and the production of our introspective judgements. Conceived in isolation, they do not amount to very much. But when conceived as integrated elements in a unified neurological framework, they evoke a radically different mode of self-apprehension.

Such examples are easily multiplied, and fascinating to explore in detail. But these will serve to illustrate the desired point, and I shall leave their multiplication and loving exploration to those whose competence in human physiological psychology is more substantial than my own. There is a further and potentially larger class of examples that want at least a mention, however, and these are of interest because they do not represent a mere reconception of states already introspected under some familiar conception or other. There appears to be a great deal about our physiological and neurological activities – activities currently opaque to us – that we can *come to* recognize introspectively, given the concepts with which to classify them and the training necessary to apply those concepts reliably in non-inferential judgements. Interesting as well is the concomitant expansion of the range of internal activities over which we can apparently gain some voluntary control (again, given suitable training). Examples here would include various aspects of the cardiovascular system – specifically, blood pressure, heart rate, capillary constriction and dilatation[7] – and various kinds of electrical activity in the CNS, most notably, the recognition and voluntary control of the so-called 'alpha rhythms' in the gross electric potential across the cerebral cortex.[8, 9] Indeed, it seems we can even learn to exercise immediate voluntary control over the electrical behaviour of single motor cells,[10] and clusters of cortical cells.[11]

Altogether then, it appears that the level of self-awareness and self-control that modern man has reached is but one stage in a long journey whose greater part has yet to be traversed, a journey in which progress is measured by the articulation of better and better conceptual (theoretical) frameworks concerning the internal dynamics of human behaviour, and by their increasingly sensitive application in the business of non-inferential self-apprehension.

That journey will not be over, of course, until we possess a framework that allows us to replace as well the notions of 'conceptual apprehension', 'non-inferential judgement', and others that have carried so much weight throughout this essay. For these notions are ultimately elements of the P-theory as well: of that part of the P-theory concerned with our specifically *cognitive* capacities. And a more penetrating framework in which to understand our cognitive abilities must surely be at the very top of our list of intellectual desiderata. Indeed, there are excellent reasons for ascribing the difficulties that currently afflict epistemology to its continued and unreflective commitment to the sentential or P-theoretic paradigm, a paradigm whose integrity has already been impugned on a variety of grounds. With this thought in mind, let us now return to the more general epistemological concerns with which this essay began.

[7] Obrist, Black, Brener, and DiCara (eds) *Cardiovascular Psychophysiology* (Chicago, 1974).
[8] Jackson Beatty, 'Learned Regulation of the Human Electroencephalogram', in *Biofeedback: Theory and Research*, ed. Schwartz and Beatty (New York, 1977), pp. 351–70.
[9] Barbara B. Brown, 'Identifying One's Own Brain Waves', ch. 11 of *New Mind, New Body* (New York, 1974), pp. 353–69.
[10] John V. Basmajian, 'Learned Control of Single Motor Units', in *Biofeedback: Theory and Research*, pp. 415–32.
[11] Fetz, E. E. and Finocchio, D. V., 'Operant conditioning of isolated activity in specific muscles and precentral cells', *Brain Research*, vol.40, no. 19 (1972).

Moreover, it will be said, if we make our concerns here exclusively normative, we may relinquish the responsibility of descriptive/ explanatory adequacy that we should otherwise have to assume. If our concerns are purely normative, we can count our goal achieved when we have formulated a set of principles that, when faithfully applied in given circumstances, systematically replicate for us the epistemic choices that an "ideally rational man" would make in those circumstances. Whether that rational man (or these, or those) actually "applied" those very principles in our envisaged set, or those in some equivalent set, or whether he "applied" any principles at all, need not be a matter of especial concern to us as purely normative epistemologists. Dynamical and explanatory matters, that is, may safely be left out of account. So long as our normative theory yields the right outputs, as specified above, the determination of the actual etiologies of those outputs as they occur in real thinkers can be left to those whose concerns are explanatory rather than normative. Dropping the requirement of explanatory adequacy, therefore, frees normative epistemology from an irrelevant burden that would otherwise cripple its development, and we can pursue the matter of rational belief change in abstraction from the matter of what hardware(s) might variously instantiate the process, and from the matter of what correlative dynamic(s) might actually govern its sundry incarnations. Our concern here, after all, is with the ideal rather than with the real.

This willingness to abstract from the nature and dynamics of human psychology is common, in some substantial degree, to almost every current philosophical approach to the problem of rationality in matters epistemological. A good example would be the tradition of modern inductive logic. Here the resources of the mathematical calculus of probability are put to work in an attempt to explicate such notions as degree of confirmation, inductive coherence, and rational belief change. But these reconstructions involve no suggestion that normal humans actually engage in the covert use of that calculus, or perform any covert mathematical operations at all, in the business of readjusting and expanding their beliefs. Similarly, other approaches recommending other principles to guide our intellectual evolution abstract unblushingly from the creative aspects of the process (the "context of discovery"), centre attention on questions of justification (criticism, evaluation), and inveigh mightily against the evils and irrelevance of "psychologism". For example, within Sir Karl Popper's approach to epistemology this process of abstraction reaches a natural conclusion in his notion of evolution within *world 3*, an objective domain consisting essentially of propositions, theories, and arguments.[1] It is in this domain, argues Popper, that scientific evolution proper takes place, and '... world 3 science can be investigated only logically'.[2]

This near-universal willingness to push questions of a descriptive/dynamical/explanatory kind to one side – the better to pursue our normative interests – would perhaps be alright if we could be certain that the descriptive parameters of our current conception of human intellectual activity capture what is really relevant to and adequate for the statement of a general normative theory. But what right have we, at the present stage of our understanding, to feel certain about any such thing? The common-sense P-theory, whose categories are taken as a given by almost all approaches to normative epistemology, is a theory whose basic integrity is very much open to doubt. And even if its portrayal of human nature suffers no faults more grievous than simple superficiality, it is still an open question whether the fundamental parameters of rationality are to be found at the categorial level it comprehends. It is my conviction that they are not to found at that level, and in this final chapter I shall try to outline some important reasons for holding such a conviction.

Generally speaking, the question I wish to pursue in this final chapter is 'In what direction has normative epistemology a future?' Given certain heterodox trends in recent epistemology, this question has become worth asking – indeed, has become rather pressing – quite independently of the considerations just outlined. Prominent among the relevant writings are those of T. S. Kuhn,[3] J. Piaget,[4] and W. V. Quine.[5] The common thrust is the idea that epistemology should be naturalized: the idea that epistemology,

---

[1] See Karl Popper, *Objective Knowledge* (Oxford, 1972). *World 3* contrasts with *world 2* (the domain of thoughts, beliefs, and other mental states), and with *world 1* (the familiar domain of physical objects and processes).
[2] Karl Popper, 'Replies to My Critics', in *The Philosophy of Karl Popper*, ed Schilpp (LaSalle, 1974), p. 1148.
[3] *The Structure of Scientific Revolutions* (Chicago, 1970).
[4] *The Child's Construction of Reality* (London, 1955). For an engaging outline of Piaget's overall view of the status of epistemology, see *Insights and Illusions of Philosophy* (New York, 1971).
[5] 'Epistemology Naturalized'.

properly conceived, is a part of *developmental psychology*, individual and social.

The intellectual motives behind such a view converge from several different directions. One can be impressed with the poverty of current *a priori* epistemology when confronted with the intricate details and the grand dramas of our actual theoretical development (cf. Kuhn). Or one can be impressed with the intriguing facts concerning intellectual development in children, a development with dimensions quite uncomprehended by orthodox epistemology (cf. Piaget). Or one can be convinced on both historical and systematic grounds that the goal of an *a priori* explication-cum-validation of the methods of science is a foolish or unreasonable one, in which case there is no reason not to make an honest woman of epistemology by wedding it to psychology straightaway (cf. Quine and Piaget). And there are other motives as well, as we shall see. But the claim that the enterprise of epistemology should be conducted along the lines of any other natural science renders problematic the status of what we would call *normative* epistemology. 'Ought's not being derivable from 'is's, it would seem that normative epistemology cannot be a purely empirical science.

At one level, the awkwardness here is perhaps easily rectified. We could (i) insist on the distinction between descriptive epistemology and normative epistemology, (ii) agree that the former is a desirable enterprise much in need of further development, (iii) confess that the normative discipline has indeed been parochial in the extreme, and (iv) resolve to conduct it in future with a more systematic eye on the rich variety of empirical facts concerning human intellectual development. And I think that such a response is not incorrect. None the less, I think it is also true that this simple act of contrition is unlikely to uproot a certain complacency with respect to the shape and direction of modern normative epistemology, a complacency that is not at all justified given certain of those very facts about human intellectual development that we have just resolved to respect. The facts, or apparent facts, that I wish to appeal to concern the earliest stages of human intellectual development; and what I wish to attack is a very broad class of theories I shall call 'sentential epistemologies'.

18. *The sentential kinematics of orthodox epistemology*

Given a modern understanding of biological man and his place in nature, it is natural to conceive of ourselves as *epistemic engines*. At the urging of a chemical energy flux and under the influence of sensory stimulation, we spend our lives proceeding through an almost continuous succession of "epistemic states", in accordance with some marvellous inner dynamic, and the process is such that, in general, later states are in some important way superior to earlier states. The elucidation of criteria for judging the relevant kind of excellence and for helping to guide our development has long been the defining task of normative epistemology. But what is the basic unit on which evaluation is to be made? Of old, it was often taken to be the individual idea, although this usually unfolded into the propositional matter of how much one's idea contained, or of how much of the idea one had grasped. Later it became the judgement or belief, and then the interrelated cluster of beliefs or large-scale theory. Most recently we have become both holistically inclined and even metrically minded. Currently, concern tends to centre on the comprehensive set of all of a thinker's beliefs at a time, on the degrees of credibility of, or credence given to, its elements, and on principles for guiding the transition from one comprehensive set of beliefs to another and/or from one credence-configuration to another.

In fact, it is difficult to find a current approach to normative epistemology that is not in some way an instance of what I shall call (with a little licence) the *ideal sentential automaton* (ISA) approach. On the ISA approach we assume that, at least for the purposes of normative theory, the current state of an epistemic engine is relevantly and adequately represented by a set of *sentences* or *propositions*, perhaps with weights assigned to each, representing probabilities, degrees of credibility, degrees of entrenchment or tenacity with which they are held, or whatever. We assume also that the epistemic system is subject to inputs, *also representable by sentences*, which can be of two distinct kinds: (i) fresh observations, and (ii) new hypotheses. The search for the essence of rationality consists, on this approach, in the search for some suitable function or relation R that maps states and inputs onto or into subsequent states – that is, into other sets of sentences with different members and/or differently assigned weights. A given transition, from one set to another, can then be counted as rational just in case it is an instance of the relation R, and a given state can be counted as rational just in case it is the result of a rational transition. (One can think of many plausible independent constraints to put on rational states – that they be consistent, that they be inductively coherent, and so forth – but these conditions can be incorporated into the definition of the relation R.) Rational intellectual development, accordingly, is represented by a long sequence of sets of sentences, suitably related as predecessors and successors in the manner described.

And the final assumption of the ISA approach, an entirely natural one given the story to this point, is that the parameters relevant to the definition of the relation R are to be found in the wealth of logical and quasi-logical properties (mostly relational) holding of and among *sentences,* or *sets* of sentences, or *sequences* of sets of sentences. *Rationality, it is assumed, is at bottom a matter of sentential parameters.*

The generality of the ideal sentential automaton approach must be appreciated. Though I permit myself the use of the term 'automaton', an instance of the ISA approach need not be committed to the idea that R must be a function (that each input/ prior-state pair has a uniquely rational successor state). Many alternative states might be equally (indifferently) rational successors to any given input/prior-state pair. R, that is, can be a relation.

Nor does the ISA approach involve any particular story on inputs, on their genesis or epistemological status. One could be anything from a sense-datum theorist to a Feyerabendian on the matter of observation judgements, and still embrace the ISA approach. And the same point holds for the intuition or generation of new hypotheses. Furthermore, beyond the crucial assumption cited at the end of the next to last paragraph, no significant constraints have been placed on the interpretation given R. Views as diverse as those of Hempel, Carnap, Popper, Bayesian subjectivists, and of most other theorists both internal and external to the logical empiricist tradition, would all count as instances of or contributions towards the ISA approach. For the crucial element is the commitment to a sentential kinematics, to the vision of rational intellectual activity as consisting essentially in a dance of propositional states, a dance whose form preserves certain propositional relations.

Plainly, the spirit of the ISA approach has dominated epistemological research in this century, and the momentum it has generated is enormous. Nor is this surprising. First, the ISA approach could hardly be more intuitive. In representing a man-at-a-time by a set of sentences, and in taking as the relevant evaluative parameters the many properties and relations holding on sentences and sets of sentences, the ISA approach is taking its cue from the plainest of common sense. One does regard oneself, and others, as a seat or subject of beliefs (convictions, suspicions, and so on), and, as with other intentional states, specific beliefs are identified by the employment of a specific *sentence*: *x* believes that *p*, we say, where some declarative sentence substitutes for the second variable. And in our everyday intellectual commerce, evaluation and criticism characteristically centre on beliefs (convictions, suspicions), and involve appeal to other beliefs and to relations holding thereon such as consistency, entailment, and so on, relations *also* holding, perhaps primarily, on the identifying *sentences*. Accordingly, the approach at issue appears as a straightforward attempt to make more systematic hay out of what is already our fundamental conception of ourselves as epistemic beings. The ISA approach, that is, has its roots squarely in the P-theory: the intellectual kinematics of the ISA approach is also the intellectual kinematics of common sense. Secondly, the approach is also appealing because the theoretical resources at hand are substantial, being the whole of modern logic and, by way of the likes of the probability calculus and information theory, much of modern mathematics as well.

Despite these favourable presumptions, however, it is the thesis of this final chapter that rational intellectual development in an epistemic engine cannot at bottom be adequately, nor even perhaps relevantly, modelled or represented by sequences of sentence-sets and by the properties of and relations between them. That is, there is significant reason to believe that the problem of reckoning just what epistemic virtue really consists in will not be solved, nor even confronted, so long as we remain transfixed within the tradition of sentential epistemologies.

19. *Continuity: the problem of the early stages*

The quickest and surest way to conceive a lasting suspicion of the ISA approach is to consider what might be called the problem of the early stages. The basic argument here is quite simple.

(1) Rational (healthy, virtuous) intellectual development in an infant cannot be accurately or even usefully represented by a sequence of sets of sentences suitably related. Bluntly, intellectual development at that stage is not ISA-representable.

(2) Rational (healthy, virtuous) intellectual development in an infant is entirely continuous with – is not different in fundamental kind from, is basically the same kind of activity as – rational intellectual development at later stages, even much later (i.e. adult) stages.

∴(3) As a general approach to what rational intellectual development consists in, the ISA approach is pursuing what must be superficial parameters. That is, sentential parameters cannot be among the primitive parameters comprehended by a truly adequate theory of rational intellectual development, and the relevance of sentential parameters must be superficial or at best derivative even in the case of fully mature language-using adults.

If this argument is sound, contemporary epistemology has been pursuing a mistaken ideal.

Before proceeding to the discussion of this argument, let me make two preliminary remarks. First, I would emphasize that this argument is intended as an attack on the ISA approach not just as an approach to descriptive or "natural" epistemology, but as an approach to normative epistemology as well. If we cannot characterize/explicate/account for *intellectual virtue* in the developing infant, then, unless intellectual virtue at later stages consists in something different from what it consists in then, we shall not be able to give an adequate account of what intellectual virtue consists in at *any* stage.

Second, though we shall find some highly persuasive considerations in support of both (1) and (2), it will not distress me overmuch if the reader judges them to be less than wholly decisive, for a more basic conundrum here will remain. Neither of (1) or (2) is a particularly subtle point, and neither of (1) or (2) is obviously false. But the popular conviction that adult rationality is essentially or primarily a matter of sentential parameters depends on the falsity of at least one of them. And at the very least it is problematic which of them must go. Altogether then, this is a situation that must be reckoned with.

Let us begin with the easier of the two, premiss (1). The behaviour of an infant during the first several months after birth invites description/explanation in terms of specific perceptions, beliefs, and reasonings no more than does the (more leisurely) behaviour of many plants, or the (more frantic) behaviour of many microscopic protozoa. The apparently chaotic economy of an infant's behaviour – and there is plenty of behaviour if one looks closely – is not rendered transparent or coherent by any such projection of familiar categories. The relevant organization in its behavioural economy has yet to develop. Were it not for the fact that infants resemble and eventually develop into thinking adults, whereas plants and protozoa do not, we would not even be much tempted to ascribe propositional attitudes and our usual cognitive concepts to them.

The truth here would seem to be as follows. As a result of sensory stimulation and its own internal activity, the infant undergoes a process of development that *eventually* produces or manifests itself in an economy of behaviour that *does* invite fruitful description and explanation in terms of the propositional attitudes, in terms of specific perceptions that p, beliefs that q, etc., and the familiar patterns of inference. But that stage is not reached until several months after birth, at least, and the developmental process that leads to it is a process we understand hardly at all. Certainly we do not understand it in terms whose eventual propriety, after all, is the *result* of that development.

Now a contrary position here is perhaps not utterly impossible. One could insist, as some will, that the young infant is none the less the subject of the familiar propositional attitudes, the claim being that the particular propositions that would express the infant's attitudes are inexpressible in our language, the infant's ideas or concepts being primitive ones quite different from our own. Our inability to make any systematic explanatory sense of their behaviour and development in terms of propositional attitudes is therefore not to be marvelled at. Thus, it will be said, one can embrace premiss (2) whole-heartedly, while rejecting (1).

Now it must be admitted that a position along these lines is in no immediate danger of empirical refutation – which perhaps explains its currency in some circles – for it is a near paradigm of an untestable hypothesis. A dispassionate look, I suggest, will reveal it for what it is: an *ad hoc* assumption of negligible testability embraced solely to preserve the generality of the 'propositional attitude' conception of human intellectual activity. One can indeed explore the simpler conceptions of language-using children, and even trace their conceptual development with some degree of assuredness (though without, it should be noted, significant success on the explanatory front). But in the case of infants of at least the first several months or so the exercise is palpably unreal. Moreover, nothing is really gained by these untestable assumptions, for the inevitable is merely postponed. Even if we are willing to ascribe inarticulable propositional attitudes to the infant, we must still reckon – at some point in the history of the foetus/infant – with the problem of a transition from a period of intellectual development *not* characterized by the manipulation of propositional attitudes to a period of intellectual development that is so characterized. The assumption at issue merely pushes the period of problematic transition an indeterminate distance further back in the infant's history, without empirical sanction, and to no explanatory or predictive advantage. The only advantage it ever seemed to have – the classical idea that our mature concepts can be analysed or decomposed into a relatively small number of simple/natural/primitive concepts, whose acquisition or possession is readily explained – has long since ceased to be plausible on either count. On the first count, our mature concepts simply do not decompose in the desired fashion, certainly not into any identifiable set of ultimate or "simple" concepts. Nor, correlatively, is conceptual growth generally explicable in terms of the increasingly complex articulation of combinations of simples. And on the second count, the classical ideas on the acquisition of these alleged simple concepts remain as sterile as they ever were, Hume's "faint copy" theory being the wholly unacceptable best of a badly mysterious lot. And the situation has deteriorated substantially since then, since it is no longer clear that there can even be such a thing as a "simple" (that is, a non-theory-laden or network independent) concept. In sum, the projections of doting mothers and desperate philosophers notwithstanding, we have neither empirical nor theoretical reasons for supposing the young infant to be the subject of propositional attitudes. At these early stages, therefore, we must conceive of the intellectual activity of the infant in some way other than that embraced by the ISA approach.[6]

Perhaps there will be wide agreement on this much, the residual differences being over just when in the developing infant's career it becomes appropriate to ascribe specific judgements, specific beliefs, and the drawing of inferences from *p*s to *q*s. What must now be argued is that for any plausible drawing of the line, development across that line marks no fundamental novelty or change in the basic kind of activity the infant is engaged in. For my own part, I do not know where a line should be drawn, though I am inclined to think of four months and twelve months as the early and late limits respectively. Even this assumption will not prove crucial, however, as I think we shall find continuity everywhere we look. Let us turn then to the question of premiss (2).

Some of my colleagues who were kind enough to read versions of this material in earlier drafts have advised me that the claim being set forth in premiss (2) is rather more opaque than it should be. Before attempting to defend it, therefore, I shall attempt to clarify it. And I think its point will emerge clearly against the backdrop provided by the following analogy.

A volume of gas, as we now appreciate, is a swarm of high-velocity particles bouncing in a perfectly elastic fashion off one another and off the walls of whatever container confines them. But one does not need to conceive of gases in this way to get a rough handle on their behaviour. One can conceive of a gas as just a compressible fluid, and use the classical gas law, $PV = \mu RT$. (Here V is the volume of the enclosed body of gas, P is the pressure it exerts on the walls that enclose it or at any point within it, T is the temperature of the gas, μ represents the mass of the enclosed gas, and R is just a constant of proportionality.) On this early classical conception of things, the temperature and pressure of a gas were each taken to be a property that has a well-defined value at any point within the gas (or against the wall), a property that varies continuously throughout the volume of the gas (or the surface of the wall) and throughout time. These early classical notions can generally be applied without difficulty, by using fluid-

---

[6] For a recent and noteworthy attempt to make a go of a linguistic interpretation of the infant's cognitive activities, see J. A. Fodor, *The Language of Thought* (New York, 1975). For an effective critique of this bold attempt, see Patricia Smith Churchland, 'Fodor on Language Learning', *Synthese*, vol. 38 (1978).

expansion thermometers and mercury-column barometers, and the classical gas law will perform moderately well given such inputs.

There is, however, an obvious difficulty with these classical notions, and it shows up most clearly if we consider a volume of extremely rarefied gas. Consider, for example, a room-sized enclosure that is completely evacuated but for five or ten oxygen molecules knocking enthusiastically about. In such a case it is plain that there is no continuous force, exerted by the gas, on any area of the enclosing walls; rather, each wall is periodically banged by a particle and suffers a discontinuous change in momentum. And a similar difficulty holds for early classical temperature: there is no property of the kind required that is well defined for and varies continuously over each point within the "body of gas" in question. If we divide the enclosure in thought into tiny sub-volumes, each of these latter spends most of its career as a hard vacuum with only the occasional fleeting visitor providing a discontinuous break from that monotony. Furthermore, with a gas so rarefied as in our example, standard thermometers and barometers will prove unequal to the task of providing trustworthy figures (or any at all), and, more importantly, even ideal instruments of that kind would yield no steady or coherent results from the point of view of the classical gas law. As can be seen then, the internal activity of an extremely rarefied gas is not accurately represented by the parameters of early classical temperature and pressure, and the classical gas law that uses them is unable to explain and predict the behaviour of such a gas.

However, if we steadily add more and more energetic oxygen molecules to those already within the enclosure, the situation with respect to the classical parameters changes steadily as well. As the volume becomes more crowded with moving particles, and the spotty hailstorm on the walls becomes a steady buzz and then a smooth hum, the situation within the enclosure approximates with increasing closeness the classical conceptions that proved so problematic earlier. As the density within the enclosure approaches familiar levels (about 10^{24} particles/m^3), even very tiny sub-volumes will always contain a representative mix of particles whizzing through them at any given time, and the rate of change of momentum to which any section of the wall is subject will be steady and representative even for very small subareas. Ther-

5

Sentential epistemologies and the natural science of epistemic engines

17. Normative epistemology: the problem in perspective

A child does not need an epistemological theory. He learns with a relentless efficiency his adult incarnation will envy, but he makes no conscious use of explicit principles to guide or shape the runaway evolution of his world picture. For the most part, of course, neither does an adult. Upon being pressed for a justification or explanation of some conviction or epistemic decision, an adult may respond with the likes of 'Well, such-and-such implies that P', or P would explain so-and-so', or 'P is the only serious possibility I can think of'. But in the vast majority of cases such humble remarks exhaust the speaker's awareness of whatever principles we might assume to govern his intellectual evolution. And yet in adults as well as children that evolution displays a richness and complexity that explanations of the sort just cited barely begin to penetrate.

That complex evolution, therefore, wants accounting for, as it occurs both in infants and in adults. We wish to understand in detail the concert of factors that produce and shape it. So far, our concerns will be purely theoretical (descriptive, explanatory). But our concerns do not end here. In particular, we wish to understand what factors and principles guide intellectual development in a *rational* man, indeed, in an ideally rational man. And here our concerns have become overtly normative: our interest is centred on the possible acquisition of an organon that will make possible the authoritative criticism of one's past intellectual conduct and the authoritative guidance of one's epistemic future. The better we understand The (epistemic) Good, that is, the closer our (epistemic) behaviour may be made to approximate it. One need not be deceived that the final goal is within easy reach, but this does not make the search for it any the less worthwhile.

mometers and barometers will behave tractably, and with the classical gas law we can give a fair to middling account of the gross behaviour of non-rarefied gases.

Just the same, it is easy to see that the behaviour of the increasingly crowded volume to which we were steadily adding particles does not *come* to be a function of certain parameters of which it was not a function initially. Strictly speaking, the body of gas never acquires the classical properties at all, though we get away with the pretence when the particle density is sufficiently high. The crucial parameters of the behaviour of the gas were and remain quite distinct from these. They are still precisely what they were in the rarefied case, and they concern the average velocity of the particles, the mass and dimensions of those particles, their angular momenta, their internal oscillations, their number for a given volume, and so forth. And a truly general and successful understanding of what a gas is, what it does, and why it does it simply cannot be had unless it is based on these more penetrating parameters, as many further empirical failures of the classical conception attest.

The point of this story is that the basic parameters of gaseous behaviour are the same whatever the degree of rarefaction of the gas, notwithstanding the increasing degree to which that behaviour becomes "comprehensible" in terms of a different set of putative parameters as particle density increases. Analogously, the claim of premiss (2) is that the basic parameters of rational intellectual activity are the same whatever its stage of development, notwithstanding the increasing degree to which that activity becomes "comprehensible" in terms of propositional attitudes as we consider the later stages of its development. And I intend that this analogy should be taken quite seriously. We already have it from the preceding chapter that the framework of propositional attitudes constitutes a theory of intellectual activity, and a noticeably imperfect theory at that. And we also have it from premiss (1) that there is a class of cases, apparently continuous with our own, where the framework of propositional attitudes fails to find descriptive/predictive/explanatory purchase. All we need now to complete the analogy is the not utterly implausible claim of premiss (2) that the case of human infants is more than just 'apparently continuous' with the case of cogitating adults. Let us see what can be said in favour of such a claim, and let us examine three dimensions of human development: the behavioural, the structural, and the functional.

Since common sense contains no systematic way of understanding the intellectual states and activities of pre-linguistic children other than in terms of perceptions, beliefs, inferences, and so on (that is, in the same terms in which we conceive of the intellectual activities of language-using adults), our tendency to so conceive of them is not at all surprising. And it is doubly unsurprising when one considers the rich behaviour of the child of about a year. Though its vocabulary consist of but a handful of words, or of none at all, such a child has a detailed conception of the world and an ability to anticipate and control his environment that is remarkable. The steady ascription of specific perceptions, beliefs, desires, and the familiar patterns of inference to him is both natural and fruitful. Claims of 'projection' will likely fall on deaf ears.

But the degree to which infant behaviour invites the ascription of specific propositional attitudes, and the degree to which such ascriptions are predictively and explanatorily fruitful, falls off continuously as we consider earlier and earlier stages within the first year, and one's awareness of the extent to which those ascriptions are lame projections or plain romancing increases with equal smoothness. One might of course go on credulously projecting several months back into the womb – no behavioural *discontinuities* bar such foolishness – and that is really the point. The steady articulation of the infant's behavioural repertoire – from a stage where our usual cognitive concepts simply fail to find adequate purchase to a stage where they find paradigmatic application – supplies no comfort at all to one who would quarrel with the continuity of development claimed in premiss (2). If there is any basic change or shift in the infant's mode of intellectual activity during that first year, it does not show itself in any characteristic change in the development of its behaviour.

There will be some who say that we are here looking for the shift in the nature of the infant's cognitive activity too early in its life. Perhaps the shift from an activity that does not consist in the interaction of propositional attitudes to an activity that is so constituted takes place coincidentally with and as a result of the child's acquisition of a language during the second and third years of its life. On this suggestion, the change in the nature of the child's cognitive activity would be reflected in his display of complex linguistic behaviour.

This suggestion fares poorly under close scrutiny. The first point to make is that this criterion for locating the alleged shift in the nature of the child's cognitive activity clashes badly with a second criterion with which *prima facie* it should cohere closely. That second criterion is the degree to which the child's continuing behaviour yields profitably to systematic explanation and prediction in terms of the familiar propositional attitudes. And the difficulty is that the child's behaviour is to a substantial degree comprehensible in such terms long before he has any significant command of language. As remarked earlier, by the time he has reached the age of a year or so, such interpretations of his behaviour are adopted by everybody. But his command of the language is probably nil, and will not amount to very much for another year or two in any case.

The second point to be made – in response to the suggestion that language acquisition marks a shift to a mode of cognitive activity distinct from the native mode – is that language is clearly *learned*. It is a complex skill whose acquisition and continued exercise must depend on the use of those cognitive and executive capacities characteristic of the pre-linguistic child. Just as walking is a learned activity, the result of a carefully orchestrated exercise of our native motor capacities, capacities whose continued exercise makes continued walking possible; so is intelligent speech the result of a carefully orchestrated exercise of our native (pre-linguistic) cognitive and intellectual capacities, capacities whose continued exercise makes continued language use possible. The so-called 'linguistic mode of intellectual activity' does not displace our native mode of intellectual activity. Rather, linguistic activity, whether overt or covert, is just one of a great many tricks our basic machinery learns to perform. To the extent that there is a specifically "linguistic" mode of intellectual activity then, it is constituted and sustained by more fundamental modes of intellectual activity.[7] So far then, premiss (2) is sustained.

When we turn to physiological or structural considerations the situation with respect to continuity appears, at least at a gross level, much the same. Though generally somewhat immature, the major

[7] This claim, it should be noted, is quite consistent with Noam Chomsky's view (*Aspects of the Theory of Syntax* (Cambridge, Mass., 1965)) that humans embody an innate structure for the learning of specifically human languages. What I do deny is that the relevant learning process is itself ISA-representable, and that such infra-linguistic activities are displaced by the linguistic activities they make possible.

brain cells or neurons are already formed some months before birth. Development from there consists most noticeably in such things as the branching growth of dendrites, the lengthening of axons, and the progressive myelinization of these interconnective impulse-carrying fibres. These activities take place at somewhat different rates in different portions of the brain, and are easily correlated (in a gross way) with certain areas of behavioural development. But all of these processes begin very early (pre-natally, in fact), develop continuously, and continue to develop through infancy, childhood, and well into adult life.[8] This apparent continuity of structural development also speaks in favour of premiss (2).

Consider finally the matter of function. Here, to be sure, we know very little, but what we do know speaks perhaps the most strongly in favour of (2). We do know that the network of neurons/axons/dendrites generates and conducts nervous impulses, and that the conductive/generative dispositions of that network can be modified by the successive impulses to which it is subject. We do know that the gross structure of that network is of a kind that fits it for the processing of information, and for evolving in the manner in which it does so. (A television set, for example, processes inform-ation when it filters out a highly specific display from the storm of radiation incident upon it.) And we know that the structural development the brain undergoes fits it for processing information in more and more complex ways. But nervous activity in a network of this kind is a characteristic of the brain at *all* stages of its develop-ment, even to some degree of its immature, pre-natal stage. The brain, in short, is a self-modifying information processor from its inception to its death, and should some portion of that information-processing capacity be devoted to the execution of linguistic routines during some substantial portion of the brain's existence, that fact appears as almost incidental from this more general point of view.

Given this, and given the continuity of behavioural and struc-tural development as well, premiss (2) acquires a decided air of plausibility. Accordingly, we must take seriously the conclusion drawn earlier. To repeat it:

As a general approach to what rational intellectual development consists in, the ISA approach is pursuing what must be superficial

[8] W. A. Marshall, *Development of the Brain* (Edinburgh, 1968).

parameters. That is, sentential parameters cannot be among the primitive parameters comprehended by an adequate theory of rational intellectual development, and the relevance of sentential parameters must be superficial or at best derivative even in the case of fully mature language-using adults.

20. *The poverty of the ISA approach: further considerations*

Upon reflection, the conclusion just reached should not seem surprising. For what reason, after all, should we take the form of the verbal chatter we exchange as being fundamental for a general theory of virtuous development in an epistemic engine? The kind of causal interaction we call language use is a dimension of activity acquired only after an enormous amount of prior cognitive development on the part of the human infant, and it is an ability that is never acquired by any or most of the many other natural epistemic engines with whose company evolution has graced us. Of our many and varied cousins among the higher animals at least, it is plain that they are all centres of cognitive activity at some level or other, and all display the capacity for learning from experience. But language plays no role in their activities. Viewed from this broader perspective, language appears as a peripheral phenomenon idiosyncratic to a single species of epistemic engine. And how much more idiosyncratic must its forms and elements appear when we consider the much greater range of epistemic creatures that the universe at large must surely contain? It is not difficult to imagine creatures, our cognitive equals or superiors, who use a systematic medium of information exchange between individuals, where the elements and internal structure of that medium bear no similarity at all to the elements and structures of any human language. (In fact, we may have such a medium right here at home in the so far impenetrable clicks and whistles of the agreeable dolphins.) In sum, the idea that the fundamental parameters of cognitive development and intellectual virtue should find themselves displayed in the structure of human language is as parochial as it is optimistic.

It may appear to some readers that the position here taken – on the relatively peripheral role that linguistic structures will play in fundamental epistemology – is inconsistent with the very substantial role ascribed to language as the bearer and shaper of cognitive/perceptual categories back in chapter 2. I shall therefore digress momentarily from the further promotion of the former position in order to show that it is not in conflict with the latter. This is best done, I think, with a sketch of the positive role that the institution of language does play in a society of epistemic engines like ourselves.

As I think the reader will grant, the sensory periphery of the human nervous system is at any given time in receipt of perfectly ridiculous amounts of information. To appreciate just how enormous the amounts are we must consider how many individual sensory cells there are at the periphery of our collected modalities (visual, auditory, tactile, olfactory, gustatory, thermal, kinesthetic, visceral), and how many discrete states are possible for each such unit. If we consider only the 300 million or so rods and cones in our visual system, and if we assume that each such unit is capable of only two states – stimulated and unstimulated – then we must conclude that the information contained in any momentary state of that system is about 300 million binary bits. And more realistic assumptions about the range of states possible for rods and cones would probably raise this figure by at least another order of magnitude. Now given that comparable calculations for the other modalities yield comparably large figures, and given that the total state-configuration of the entire sensory periphery is continually changing from split second to split second, one begins to gain an appreciation of what the brain is up against, literally as well as figuratively.

Clearly it is not information at this unprocessed/indiscriminate level that is exchanged when we use language. It is effectively impossible to exchange all of it, and the great bulk of it is useless in any case. But these state-configurations have interesting causal consequences, from the sensory periphery inwards along the pathways of the central nervous system and into the hierarchical labyrinth of the brain. By means of these causal processes the initial information is filtered, interpolated, amplified, integrated – over time as well as over other dimensions – refiltered, reintegrated, and refiltered again and again according to principles we have barely begun to penetrate. But though our understanding here is slim, I think it is safe to say that the brain has abstracted quite ruthlessly from the sea of information in its peripheral state-configurations when it arrives at an internal state that would be expressed by an English sentence such as 'There is a fly in my soup', or 'Dark clouds mean rain', or '$PV = \mu RT$'. Clearly, human language is a device for exchanging, among individual information-processors, information at a very high level of abstraction relative to the actual state-configurations, both past and present, of the sensory peripheries of the speaker and hearer. That is to say, by the use of language we are exchanging information at or towards the *top* of our information-processing hierarchies. Information exchange is of course maximally useful at this level, and between individuals who are sufficiently similar in the relevant respects, it is maximally feasible here as well.

To acquire a specific language is, among other things, to learn to process peripheral information into the categories that language provides. The causal processes initiated at the periphery come to terminate in the utterance of specific sentences within that language, or if not in overt utterances, then in covert states involving dispositions to assent to them. To acquire a specific language then is indeed to come to share in a specific view of reality. The informational matrix the language embodies comes to shape the processing of peripheral information "from the top down", as it were, and by this means all who learn it are acculturated into a common view of the general nature of reality and of the possible configurations it may here and there assume. Because of this agreement, and because of their common command of a publicly manipulable matrix for specifying possible reality-configurations (relative to that matrix, of course), the commerce of information-exchange in such a group can be as vigorous as we know it to be.

This then, in sketch, is the primary role that language plays among human epistemic engines. And this role is essentially that ascribed to it in chapters 2 and 3. However, it is evident that this sketch is also consistent with the criticisms of the ISA approach so far entertained – the *continuity objection* levelled in the preceding section, and the *parochiality objection* tendered at the outset of the present section. That language should play the role sketched above is quite consistent with the fact that its use must be acquired, sustained, and administered by more fundamental information-processing systems of a non-linguistic kind. And it is consistent as well with the fact that language is idiosyncratic to the human epistemic engine, and with the fact that the particular elements and structures it happens to use are not uniquely essential to its abstract function as a medium of information-exchange in any case.

While we are recalling the themes of chapters 2 and 3, there is a third objection to be considered, which we might call the *plasticity objection*. The difficulties for the ISA approach here stem from the fact that one of the most significant features of long-term and/or large-scale intellectual development in adult society is the evolution of language itself. Languages appear to be sufficiently plastic in their actual and potential responses to long-term sensory stimulation that it is highly problematic whether we can specify any permanent features or dimensions common to all languages at all times – even all *human* languages at all times – where those features or dimensions are suitable for framing a general characterization of epistemic rationality. Let me try to outline the principal difficulties in this area.

They begin with the fact that there is no fixed or neutral class of observation judgements upon whose authority all other hypotheses must be erected, or against which all other hypotheses can be authoritatively tested. There are possible an infinite number of mutually incommensurable observation vocabularies, and each presupposes its own network of speculative assumptions whose epistemic status is as problematic as the assumptions of any other theory. Orthodox empiricist versions of the ISA approach are therefore undercut completely. Observation beliefs themselves form an especially interesting *part* of the problem of rational choice between epistemic alternatives; they do not provide its solution. They appeared to do so initially only because the attention of philosophers was confined to epistemic alternatives of a minimally divergent kind.

If we are to remain faithful to the ISA approach then, it appears we must resign ourselves to comparing the relative *internal* virtues (as defined in terms of sentential parameters) of global or comprehensive epistemic alternatives. And in particular, we must be prepared to make such comparisons work in the case of mutually incommensurable global alternatives, alternatives containing different observation sentences reflecting the different calibrations imposed on our sensory modalities within each alternative. At stake in such cases is more than the question of which of two sets of sentences to embrace; at stake as well is a choice between two distinct modes of extracting information from the causal impingements of the environment.

Consequently, of the two traditional pillars of empirical

knowledge – a stable and universal experience, and a stable and universal logic – the former must be acknowledged an illusion. We are left, therefore, with the second pillar, and with the task of constructing some kind of coherence theory of rational belief, where coherence is to be characterized in terms of the stable and universal logic that remains to us.

But is not this second pillar an illusion as well? Is it not already close to being clear that the most general elements of our comprehensive belief set – the principles of deductive logic – are themselves subject to critical evaluation and rational change? Given the systematic and empirically sensitive nature of all knowledge, as discussed in chapter 3, the propositions of formal logic are epistemologically continuous with propositions of any other sort, and are therefore subject to possible modification as further elements in the general swim of rational activity. Their systemic importance, though enormous, is not infinite. For a brief outline of some relatively small-scale issues and modifications, see Aune.[9] And for an outline of some possible modifications on a larger scale, see Putnam[10] and Finkelstein.[11] The concern of these latter two papers is the potential impact of the quantum theory on the formal principles to which we currently adhere. It is not clear that the quantum theory actually constitutes a real occasion for yielding one or more of the classical principles, but the vulnerability of those principles to systematic considerations is fairly obvious. And if those principles are subject to modification in the course of rational intellectual activity, then the patterns they impose on the internal structure of one's comprehensive belief set cannot provide the fundamental constraints on what does and what does not constitute rational intellectual activity. Those constraints are to be found at a deeper level still, and if we are ever to understand what virtuous intellectual activity consists in, we must try to penetrate to that deeper intellectual kinematics of which our manipulation of sentences is just the occasional and superficial reflection.

It may seem to some readers that this "rejection" of both logic and experience leaves us no alternative but to embrace a

9 Bruce Aune, *Rationalism, Empiricism, and Pragmatism* (New York, 1970), pp. 111–115.
10 Hilary Putnam, 'Is Logic Empirical?', in *Boston Studies in the Philosophy of Science*, vol. 5, ed. Cohen and Wartofsky (Dordrecht, 1969).
11 David Finkelstein, 'Matter, Space, and Logic', in the same volume as the Putnam article just cited.

nihilism or intellectual anarchism. And indeed one prominent contributor to the debate appears to have been led by similar considerations to a position much along these lines.[12] But nothing could be further from my own purpose. I have no doubt whatever that all of a man's knowledge is grounded in the causal effects of the environment upon his earlier stages, that rational cognitive development is subject to certain general or formal constraints, that these constraints are progressively better knowable by man himself, and that such knowledge can significantly enhance his epistemic performance. What I have been objecting to, in this section and in the last, is the idea that those causal effects and general constraints can ultimately be construed *sententially*.

21. *Other horizons*

If the sentential approach is ultimately a dry well, what other approach should we pursue? If it is going to have any chance at lasting success, I suggest, it must be inextricably bound up with a deeper conception of ourselves provided by some natural science of epistemic engines generally. As to what such a science might look like I can here hazard only a few hopeful comments, exploiting ideas already in the air.

Most natural items are capable of a range of possible internal states, which states are a more-or-less determinate function of external causal factors. Accordingly, many natural items end up containing information about their environment in the simple sense that they would not (or would not likely) be in their specific present state unless the environment had manifested certain specific events, features, or patterns during the tenure of the item in that environment. Crudely, the ravages of time often leave highly characteristic marks. The growth rings in a tree trunk constitute a record of rainfall and sunshine cycles, and of parasite infestations, but not of planetary positions; a beach will hold (temporarily) a record of what walked over it, but not of what flew over it; and so on. Obviously, physical systems vary widely in their potentialities and properties as informational reservoirs. Most obviously, there is the matter of range of sensitivity. Rated as an informational

[12] Paul Feyerabend, 'Against Method: Outline of an Anarchistic Theory of Knowledge', *Minnesota Studies in the Philosophy of Science*, vol. 4; ed. Radner and Winokur (Minneapolis, 1970).

reservoir, a good hard stone scores rather poorly, whereas an operating colour/sound motion-picture camera scores extremely well, sufficiently well to suggest the metaphor, "informational *sponge*". Plainly, the animal kingdom is constituted by a great variety of informational sponges. One need only suppose the overt behaviour of such informational sponges to be a systematic function of their information-bearing states to have outlined a conception of the internal activities of natural fauna that owes nothing to our usual cognitive concepts, and which places us on a continuum with animals, trees, and ultimately even beaches.

But this conception can be only the merest first step, since an informational sponge is not yet an epistemic engine, even if its behaviour happens to be a survival-promoting function of its information-bearing states. An epistemic engine, in the sense required here, must not only have "sensory" states, and perhaps react to them; it must *learn* from them. What sense can we give to the term 'learn' within the sort of framework we are here exploring?

Here we need to imagine a certain plasticity in the functional relations holding between sensory inputs and motor outputs. In particular, we need to consider a case in which those functional relations change as a more-or-less determinate function of certain sensory consequences of their past operations. A given kind of motor response can be made more/less likely to occur again (given the input that originally prompted it) as a function of certain positive/negative "reinforcing" inputs that said response may elicit. (Pleasure and pain are familiar instances of this abstract notion.) What this will produce is a *sequence* of functional relations betwixt input and output, a sequence wherein the modifications from element to element are themselves the output of a second-order function, a function whose inputs are actualized stimulus–response pairs from the first function, plus whatever "reinforcing" states their actualization elicits.

A system of this kind will have its behaviour conditioned by the environment, and in highly specific directions. Its dispositions to behaviour will change in ways that tend to maximize/minimize the relevant occurrence of those motor responses that elicit the positively/negatively reinforcing sensory inputs. And if the occurrence of those reinforcing inputs is nomically correlated with specific kinds of circumstances, such as the ingestion of nourishment or the destruction of peripheral tissue, then we can also say that the system's behaviour changes in ways that tend to maximize the ingestion of nourishment and minimize the destruction of tissue. We have a system, that is, that *learns* to find food and avoid bodily damage. But its development in this regard is not to be explained in terms of the logical and quasi-logical relations holding of and among sets or sequences of essentially propositional states. It is to be explained in terms of the *causal* relations holding among a system of states which contain specific information as a matter of causal, nomological, and evolutionary fact.

It is important to appreciate that it is not the increasingly *adaptive* behaviour, *per se*, that makes appropriate the description 'learns from experience'. For we can readily imagine an unfortunate creature in whom the "pain" input (or other reinforcer keyed to bodily damage) has the role of a *positive* reinforcer for those contextually produced motor responses that elicit it. The behaviour of such a creature is no less conditioned by the environment than it is in the case of more fortunate creatures, and the development involved is no less a case of learning. But what that creature learns are the techniques of its own destruction.

What is common to both kinds of cases, and what is essential, I think, to any process that is to be described as learning, is the exploitation of information already in hand in such a way as to discriminate hitherto unused information among the sensory effects of the environment – in such a way, that is, as to increase the general information about the environment contained in one's non-peripheral states and behavioural dispositions. In the two cases described, prior to any conditioning, only the hard-wired system for the production of reinforcing states carried any general (or provided any specific) information concerning damaging environmental circumstances. But as a result of conditioning, a process that exploits that information, the plastic system of behavioural dispositions now contains general (and provides specific) information concerning perilous circumstances that was not contained in that system at the outset.

I am unable to sketch a general theory of the use of information to get more information, but the following example will represent a large class of cases in which this process of exploitation is iterative, in which the dividends of invested information can be successfully reinvested. What I wish to evoke here is the idea of a world that presents an *apparently* chaotic face because that face results from the

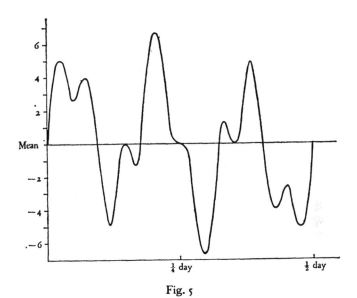

Fig. 5

interaction of many highly regular but mutually independent variables. And what I shall sketch is the manner in which that hidden order may be recovered.

Consider an intertidal environment, on an alien planet, where the graphic representation of water level against time is as in fig. 5. The survival problem confronting a creature living there is, let us say, to anticipate the variations in water level that characterize its locale (the reasons need not concern us here). Since the short-term future behaviour of tidal levels is effectively impossible to predict from short-term past behaviour, a more promising strategy for our creature to pursue would be the conditioning of some behaviour-controlling internal parameter so that the graph of its continuing ups and downs comes to mirror that of the tidal variations at issue. Let us suppose it to possess such a plastic parameter whose possible stable states consist in sine-wave oscillations, oscillations whose frequency and amplitude are subject to conditioning in the manner discussed above.

Now it is plain that no simple sine-wave regime will mimic the tidal variations perfectly, or even very closely, but a regime of $f = 6$ cycles per day and $a = 4$ units will receive stronger positive

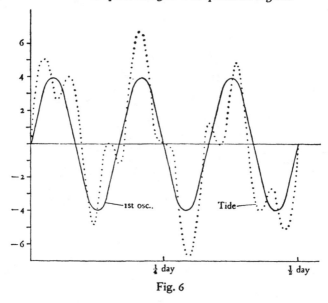

Fig. 6

and/or less negative reinforcement than any other, and we may suppose that the creature fairly quickly settles into such a regime (see fig. 6). Once there, however, new possibilities for conditioning present themselves. For the problem that now confronts the creature is the residual but occasionally serious *differences* between its anticipations and the actual tidal levels it discovers. But against the background of its adopted regime, the graph of those differences in fact is as in fig. 7. Plainly this is a problem that will yield to the same strategy invoked above. What the creature needs is a *second* oscillatory parameter, plastic with respect to frequency and amplitude, whose outputs are simply added to those of the first parameter so far as the control of behaviour is concerned, and whose long-term behaviour is conditioned in the same way and by the same reinforcers as established the first-level regime. As before, no simple sine-wave oscillation in this second parameter will mimic the relevant deviations perfectly, but a regime of $f = 16$ cycles per day and $a = 2$ units will be selectively reinforced, and we may assume that the second parameter settles into this regime. The combined output of these two parameters will then have a graph as in fig. 8, which is a substantial improvement over the singular regime of fig. 6.

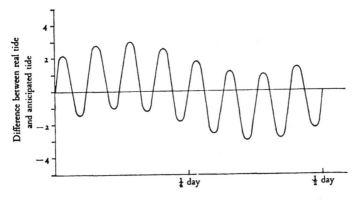

Fig. 7

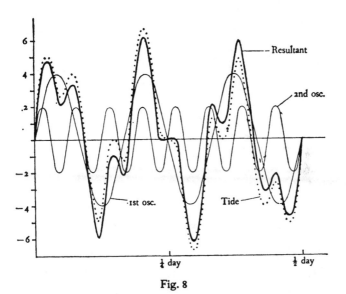

Fig. 8

Even so, the fit is still imperfect, and the graph of its deviation from reality will in fact be as in fig. 9. Repeat the story again with the conditioning of a third parameter to an oscillatory regime of $f = 2$ cycles per day and $a = 1$ unit. Here at last a sine wave does represent the deviation perfectly, and so our creature's tidal education is complete. The combined output of its three internal

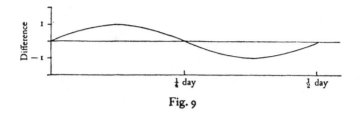

Fig. 9

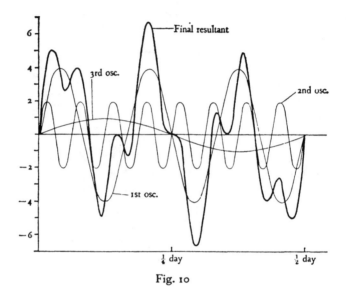

Fig. 10

oscillations will mimic/anticipate to perfection the sundry ups and downs of its watery environment (see fig. 10).

Nor need that perfection seem surprising, for the tidal patterns are produced, we may assume, by one large moon orbiting the planet three times per (longish) day, thus raising six tides per day; by a much smaller and closer moon completing eight orbits in a day, thus raising sixteen tides; and by the distant primary, relative to which the planet rotates once daily, thus raising two tides per day. The tides are in fact raised by the additive effects of three independent forces, forces which vary in precisely the fashion of our creature's three internal parameters. We might even say that our creature has learned some astronomy. Certainly its internal states contain information in that regard.

An example of this kind is instructive for a variety of reasons, I think. First, to repeat the rationale that introduced it, the example illustrates clearly the way in which information can emerge from the background "noise" in which it is buried once the more prominent regularities in that "noise" have been discriminated and subtracted from the incoming signal. It illustrates how the acquisition of some information can be parlayed into the acquisition of a good deal more. Second, although this is really the same point in a different guise, the example illustrates how the manner in which an organism processes incoming information (sensory and reinforcement information) can change as a function of the outcome of prior processing, and change to the organism's informational advantage. Third, it constitutes a case of an organism's constructing an "internal environment" that mimics/anticipates the behaviour of its external environment. Such an internal environment can function not only as a direct behavioural control, but also, as Daniel Dennett has explained,[13] as a source of reinforcement for/against other patterns of behaviour, a source of reinforcement that differs from the external environment in being non-lethal to the organism. This is an important feature of the example, since the construction of an internal "model" that can systematically mime sundry dimensions of the environment is presumably the essence of our own intellectual development.

Fourth, in our tidal creature's development we can see, recreated in miniature, the phenomenon of paradigm articulation, and of cumulative tradition, and even the possibility of intellectual revolution. The first oscillatory regime functions as the basic informational framework within which, and largely because of which, subsequent informational progress is achieved. Within the framework of a well-chosen first regime, the iterative process of winnowing out ever more subtle information can continue for indefinite lengths of time. On the other hand, the primary regime that receives the strongest initial reinforcement may turn out in the long run *not* to be the most revelatory of the subtler regularities. It may turn out that the residual deviations from reality start to get larger and more chaotic again after the nth oscillatory parameter is put in place. And it may be that no regime of which the organism is capable will

[13] Daniel C. Dennett, 'Why the Law of Effect will not Go Away', *Journal of the Theory of Social Behaviour*, 5 (1978), 2. Reprinted in Dennett, *Brainstorms* (Montgomery, Vermont; 1978).

thereafter find sufficient reinforcement to stick. Faced with chronic anomalies along these lines, what our tidal creature needs is a hard-wired system for reacting to such crises, where the reaction consists in the dismantling of whatever hierarchy of oscillatory regimes is already in place. The creature can then begin from scratch with a new basic regime of oscillation, a regime that may allow it to penetrate reality more deeply than did the basic regime it has just overthrown.[14] Ultimately, of course, the interference pattern confronting our creature may defeat even its best efforts. All of which illustrates the general point that in the business of penetrating the structure of reality, you need not only to be good, you need also to be a bit lucky.

Finally, and most importantly, the example does all this, and more, without invoking any concepts characteristic of the ISA approach.

This sketch of a non-sentential approach to epistemology is admittedly brief, but my concern here is merely to underscore its possibility.[15] Even this much, however, is sufficient to suggest some of the very different issues that an appropriately refocussed normative epistemology might confront. The virtues and short-comings of various highly general processing strategies, the resolution of conflicting informational "goals", the characterizing of an "ideal epistemic engine" (if such a unique characterization exists; it need not) – all these matters invite systematic attack. If we can come to understand much more deeply what it is that the machine between our ears is *doing*, then may we find ourselves able to specify what doing it *best* amounts to. Normative epistemology has a rich future, but it is a future that will be discontinuous with its past.

How is the evaluation of beliefs – represented by sentences – going to fit into all this? It is impossible to say in advance. Apart from the obvious fact that they render the process (partially) collective and (partially) inheritable, how representational media like human language figure in the overall process of pulling information from the environment is something that remains to be

---

[14] Cf. Kuhn, *Scientific Revolutions*. Also, Imre Lakatos, 'Falsification and the Methodology of Scientific Research Programmes', in *Criticism and the Growth of Knowledge*, ed. Lakatos and Musgrave (Cambridge, 1970).

[15] For an evocative sketch of a non-traditional approach to epistemology, and a useful bibliography in the same vein, see Donald Campbell, 'Evolutionary Epistemology', in *The Philosophy of Karl Popper*, book 1, ed. Schilpp (LaSalle, 1974).

understood. No doubt we will continue to talk, and to evaluate and criticize our linguistic behaviour, but that evaluation will eventually be located within a context that transcends the narrowly linguistic parameters of current practice. In the meantime, of course, we must proceed as usual, evaluating our epistemic activities within the unaugmented sentential framework at issue, in hopes of boot-strapping ourselves (remember the tidal creature) into a more penetrating framework of evaluation and criticism.

And finally, it appears likely that the thermodynamics of "irreversible" processes will provide the underlying conceptual framework (the "first regime") for whatever genuine progress gets made here. For it is this theory that renders physically intel-ligible such things as the process of synthetic evolution in general, and the Sun-urged growth of a rose in particular. And what is human knowledge but a cortically embodied flower, fanned likewise into existence by the ambient flux of energy and information?

Bibliography

Armstrong, D. M., *A Materialist Theory of the Mind* (London, 1968).
—— *Belief, Truth and Knowledge* (Cambridge, 1973).
Aune, Bruce, *Knowledge, Mind, and Nature* (New York, 1967).
—— *Rationalism, Empiricism, and Pragmatism* (New York, 1970).
Basmajian, John V., 'Learned Control of single motor units', *Biofeedback: Theory and Research*, ed. Schwartz and Beatty (New York, 1977).
Beatty, Jackson, 'Learned Regulation of the Human Electroencephalogram', *Biofeedback: Theory and Research*, ed. Schwartz and Beatty (New York, 1977).
Brown, Barbara B., *New Mind, New Body* (New York, 1974).
Campbell, Donald, 'Evolutionary Epistemology', *The Philosophy of Karl Popper*, ed. Schilpp (LaSalle, 1974).
Carnap, Rudolf, 'Empiricism, Semantics, and Ontology', *Meaning and Necessity*, 2nd edn (Chicago, 1956).
Chomsky, Noam, *Aspects of the Theory of Syntax* (Cambridge, Mass., 1965).
Churchland, Paul M., 'The Logical Character of Action Explanations', *Philosophical Review*, vol. 79, no. 2 (1970).
—— 'Two Grades of Evidential Bias', *Philosophy of Science*, vol. 42, no. 3 (1975).
—— 'Karl Popper's Philosophy of Science', *Canadian Journal of Philosophy*, vol. 5, no. 1 (1975).
Churchland, Patricia Smith, 'Fodor on Language Learning', *Synthese*, vol. 38 (1978).
—— 'Language, Thought, and Information Processing', *Noûs* (in the press, 1979).
Davidson, Donald, 'On the Very Idea of a Conceptual Scheme', *Proceedings and Addresses of the American Philosophical Association*, vol. 47 (1973–74).
—— 'Belief and the Basis of Meaning', *Synthese*, vol. 27 (1974).
—— 'Thought and Talk', *Mind and Language*, ed. Guttenplan (Oxford, 1975).
Dennett, Daniel C., 'Why the Law of Effect will not Go Away', *Journal of the Theory of Social Behaviour*, vol. 5, no. 2 (1978). Reprinted in Dennett, *Brainstorms*.
—— *Content and Consciousness* (New York, 1969).
—— *Brainstorms* (Montgomery, Vermont, 1978).
—— 'Brain Writing and Mind Reading', Minnesota Studies in the Philosophy of Science, vol. 7, ed. Gunderson (Minneapolis, 1975). Reprinted in Dennett, *Brainstorms*.

Dretske, Fred, *Seeing and Knowing* (London, 1969).

Fetz, E. E. and Finocchio, D. V., 'Operant conditioning of isolated activity in specific muscles and precentral cells', *Brain Research*, vol. 40, no. 91 (1972).

Feyerabend, Paul K., 'An Attempt at a Realistic Interpretation of Experience', *Proceedings of the Aristotelian Society*, new ser. (1958).

—— 'Explanation, Reduction, and Empiricism', *Minnesota Studies in the Philosophy of Science*, vol. 3, ed. Feigl and Maxwell (Minneapolis, 1962).

—— 'Materialism and the Mind–Body Problem', *Review of Metaphysics*, vol. 17 (1963).

—— 'Mental Events and the Brain', *Journal of Philosohpy*, vol. 60, no. 11 (1963).

—— 'Reply to Criticism', *Boston Studies in the Philosophy of Science*, vol. 2, ed. Cohen and Wartofsky (New York, 1965).

—— 'Problems of Empiricism', *Beyond the Edge of Certainty*, ed. Colodny (New Jersey, 1965).

—— 'Science Without Experience', *Journal of Philosophy*, vol. 66, no. 22 (1969).

—— 'Problems of Empiricism, part II', *The Nature and Function of Scientific Theories*, ed. Colodny (Pittsburgh, 1970).

—— 'Against Method: Outline of an Anarchistic Theory of Knowledge', *Minnesota Studies in the Philosophy of Science*, vol. 4, ed. Radner and Winokur (Minneapolis, 1970).

Finkelstein, David, 'Matter, Space, and Logic', *Boston Studies in the Philosophy of Science*, vol. 5, ed. Cohen and Wartofsky (Dordrecht, 1969).

Fodor, J. A., *The Language of Thought* (New York, 1975).

Hanson, N. R., *Patterns of Discovery* (Cambridge, 1958).

Hesse, Mary, 'Is There an Independent Observation Language?', *The Nature and Function of Scientific Theories*, ed. Colodny (Pittsburgh, 1970).

Hooker, C. A., 'The Philosophical Ramifications of the Information-Processing Approach to the Mind-Brain', *Philosophy and Phenomenological Research*, vol. 36 (1975).

Kant, Immanuel, *Critique of Pure Reason* (London, 1964).

Kuhn, Thomas, *The Structure of Scientific Revolutions* (Chicago, 1962).

Lakatos, Imre, 'Falsification and the Methodology of Scientific Research Programmes', *Criticism and the Growth of Knowledge*, ed. Lakatos and Musgrave (Cambridge, 1970).

—— *Proofs and Refutations: The Logic of Mathematical Discovery* (Cambridge, 1976).

Lakatos, Imre, and Musgrave, A. (eds.), *Criticism and the Growth of Knowledge* (Cambridge, 1970).

Lewis, C. I., 'A Pragmatic Conception of the A Priori', *Journal of Philosophy*, vol. 20 (1923). Reprinted in *Meaning and Knowledge*, ed. Nagel and Brandt (New York, 1965).

Marshall, W. A., *Development of the Brain* (Edinburgh, 1968).

Nagel, Ernest, *The Structure of Science* (New York, 1961).

Obrist, Black, Brener, and DiCara (eds.), *Cardiovascular Psychophysiology* (Chicago, 1974).

Piaget, J., *The Child's Construction of Reality* (London, 1955).
—— *Insights and Illusions of Philosophy* (New York, 1971).
Popper, Karl, *The Logic of Scientific Discovery* (New York, 1959).
—— *Conjectures and Refutations: The Growth of Scientific Knowledge* (New York, 1962).
—— *Objective Knowledge* (Oxford, 1972).
—— 'Replies to My Critics', *The Philosophy of Karl Popper*, book 2, ed. Schilpp (LaSalle, 1974).
Putnam, Hilary, 'Minds and Machines', *Dimensions of Mind*, ed. Hook (New York, 1960). Reprinted in Putnam, *Mind*.
—— 'The Analytic and the Synthetic', *Minnesota Studies in the Philosophy of Science*, vol. 3, ed. Feigl and Maxwell (Minneapolis, 1962). Reprinted in Putnam, *Mind*.
—— 'Robots: Machines or Artificially Created Life?', *Journal of Philosophy*, vol. 61, no. 21 (1964). Reprinted in Putnam, *Mind*.
—— 'The Mental Life of Some Machines', *Intentionality, Minds, and Perception*, ed. Castañeda (Detroit, 1967). Reprinted in Putnam, *Mind*.
—— 'Is Logic Empirical?', *Boston Studies in the Philosophy of science*, vol. 5, ed. Cohen and Wartofsky (Dordrecht, 1969).
—— 'The Meaning of "Meaning"', *Minnesota Studies in the Philosophy of Science*, vol. 7, ed. Gunderson (Minneapolis, 1975). Reprinted in Putnam, *Mind*.
Putnam, Hilary, *Mind, Language and Reality: Philosophical Papers, Volume 2* (Cambridge, 1975).
Quine, W. V., 'Two Dogmas of Empiricism', *The Philosophical Review*, 60 (1951). Reprinted in Quine, W. V., *From a Logical Point of View* (New York, 1963).
—— *Word and Object* (Cambridge, Mass., 1960).
—— 'Epistemology Naturalized', *Ontological Relativity and Other Essays* (New York, 1969).
Rorty, Richard, 'Mind–Body Identity, Privacy, and Categories', *Review of Metaphysics*, vol. 1 (1965).
Rosenberg, Jay F., *Linguistic Representation* (Dordrecht, 1974).
Sellars, Wilfrid, 'Empiricism and the Philosophy of Mind', *Minnesota Studies in the Philosophy of Science*, vol. 1, ed. Feigl and Scriven (Minneapolis, 1956). Reprinted in Sellars, *Science, Perception and Reality*.
—— *Science, Perception, and Reality* (London, 1963).
—— 'The Identity Approach to the Mind–Body Problem', *Boston Studies in the Philosophy of Science*, vol. 2, ed. Cohen and Wartofsky (New York, 1965).
—— 'Scientific Realism or Irenic Instrumentalism', *Boston Studies in the Philosophy of Science*, vol. 2, ed. Cohen and Wartofsky (New York, 1965). Reprinted in Sellars, *Philosophical Perspectives*.
—— *Philosophical Perspectives* (Springfield, Illinois, 1967).
—— *Science and Metaphysics* (London, 1968).
Toulmin, Stephen, *Human Understanding*, vol. 1 (Princeton, 1972).

Index